PERGAMON INTERNATIONAL LIBR
of Science, Technology, Engineering and Social Stud
The 1000-volume original paperback library in aid of
industrial training and the enjoyment of leisure
Publisher: Robert Maxwell, M.C.

D1454259

Quantitative and Statistical Approaches to Geography

A PRACTICAL MANUAL

THE PERGAMON TEXTBOOK
INSPECTION COPY SERVICE

An inspection copy of any book published in the Pergamon International Library will gladly be sent to academic staff without obligation for their consideration for course adoption or recommendation. Copies may be retained for a period of 60 days from receipt and returned if not suitable. When a particular title is adopted or recommended for adoption for class use and the recommendation results in a sale of 12 or more copies, the inspection copy may be retained with our compliments. The Publishers will be pleased to receive suggestions for revised editions and new titles to be published in this important International Library.

Quantitative and Statistical Approaches to Geography

A PRACTICAL MANUAL

JOHN A. MATTHEWS
University College, Cardiff

PERGAMON PRESS
OXFORD · NEW YORK · TORONTO · SYDNEY · PARIS · FRANKFURT

U.K.	Pergamon Press Ltd., Headington Hill Hall, Oxford OX3 0BW, England
U.S.A.	Pergamon Press Inc., Maxwell House, Fairview Park, Elmsford, New York 10523, U.S.A.
CANADA	Pergamon Press Canada Ltd., Suite 104, 150 Consumers Rd., Willowdale, Ontario M2J 1P9, Canada
AUSTRALIA	Pergamon Press (Aust.) Pty. Ltd., P.O. Box 544, Potts Point, N.S.W. 2011, Australia
FRANCE	Pergamon Press SARL, 24 rue des Ecoles, 75240 Paris, Cedex 05, France
FEDERAL REPUBLIC OF GERMANY	Pergamon Press GmbH, 6242 Kronberg-Taunus, Hammerweg 6, Federal Republic of Germany

First edition 1981

British Library Cataloguing in Publication Data
Matthews, John A
Quantitative and statistical approaches to
geography – (Pergamon Oxford geographies).
1. Geography – Statistical methods
I. Title
910'.01 82   G70.3   80–40911

ISBN 0–08–024296–0 (Hardcover)
ISBN 0–08–024295–2 (Flexicover)

Printed in Great Britain by A. Wheaton & Co. Ltd., Exeter

This manual is dedicated to

(i) Wally with congratulations;

(ii) Wyn with commiserations;

(iii) the establishment of a full Department of Geography at University College, Cardiff within the next six years.

Preface

THIS manual is a practical introduction to some quantitative and statistical techniques of use to geographers and related scientists. It is *not* a textbook. Each chapter begins with an outline of the purpose and necessary mechanics of a technique or group of techniques and is concluded with the most important feature of the manual, namely the exercises and the particular approach adopted. Each exercise involves an in-depth treatment of a topic of interest to geographers, and encourages the critical assessment of techniques in a context. The intention has been to emphasize the whole approach to problem-solving, rather than merely the techniques themselves. In this way, a superficial treatment of problems is avoided and students are not led to expect easy answers. The overall aim of this kind of exercise is to enhance the student's ability to use the techniques as part of the process by which sound judgements are made according to scientific standards while tackling complex problems. To this end, real data are used in the exercises (with the exception of the standard deviations in exercise 7, which were invented in order to include this interesting problem).

The manual has been written with first-year undergraduates in mind and assumes no previous knowledge of the techniques – hence the need for the outline of each technique before beginning an exercise. Graphical explanations are used wherever possible and formulae are explained in words as well as numbers. This is essential for students of Geography with an 'Arts' background; and speaking as a scientist who still requires graphical explanations to overcome mathematical 'blockages', it can do 'Science' students no harm either. Further information on the mathematics and statistical theory of the techniques must be sought elsewhere, for example in the many new textbooks covering various aspects of Quantitative Geography.

Answers are provided to numerical questions where appropriate, but many questions do not have a single, correct answer. I consider it essential that students are exposed to such uncertainties at an early date, and that they should be capable of recognizing and understanding limitations of various sorts present in all geographical work. Indeed, it is envisaged that, drawing on their geographical experience (no matter how limited), students will put forward more than one interpretation or hypothesis and that alternatives will be evaluated in discussion.

Most of the exercises began life in Geography practical classes at the University of Edinburgh and at University College, Cardiff, and have been found eminently suitable for use as class exercises. Many of the longer exercises should be divided among the students in a class and the results pooled for class discussion. Mixed-ability classes are catered for by permitting answers at more than one level of understanding, and open-ended questions towards the end of each exercise enable the student to pursue the problem to the limits of his or her ability.

Two chapters do not follow the general format. The first chapter is an introduction. In particular it provides an explicit statement of why quantitative and statistical techniques are an important part of Geography, and a framework for viewing the techniques that have been included in the manual. Chapter 15 is a conclusion. It emphasizes: (i) the choice of a suitable technique for a problem; and (ii) the collective limitations of statistical techniques. The final chapter thus provides an overview with hindsight, advising caution to those intending to make further use of the techniques.

It is hoped that this manual will increase the effectiveness of the teaching of quantitative and statistical techniques and that it will generate interest on the part of students in this part of the Geography curriculum. There is certainly considerable scope for the approach adopted here in the teaching of undergraduates, a branch of education in which the lecture is pre-eminent and perhaps too deeply entrenched. The manual and the approach are particularly appropriate as a means of introducing closely supervised project-style teaching as a preliminary to much more loosely supervised individual projects in later years of the Geography course.

Llandaff   J. A. M.
June 1979

Acknowledgements

MOST of the exercises and examples are based on data gleaned from the literature. Sources are given in the text or on the diagrams and are listed under 'References' at the end of the manual. Thanks are also due to Dr. C. K. Ballantyne, Dr. P. J. Beckett, Dr. A. Dawson, Dr. C. Harris, Mr. R. L. Hodgart, Dr. T. C. Musson, Mr. A. Reid, and particularly Dr. R. A. Shakesby, all of whom provided unpublished material and/or suggestions and comments on exercises, and to the students of Geography at Edinburgh and Cardiff who have attempted many of the exercises and have made diverse comments about them.

I would also like to record my thanks to Dr. L. Collins, Professor J. T. Coppock, Mr. R. L. Hodgart and Dr. J. A. T. Young, who, in various ways, gave me the opportunity to begin the development of the contents of this manual while I was a University Demonstrator at Edinburgh. Since moving to Cardiff, Dr. Charles Harris, Dr. Tim Musson and Professor Mike Brooks have 'allowed' me to 'take over' the Geography practical classes and have provided a very congenial working environment, in which the manual was completed. Louise Whittle was a great help in the final stages, particularly by assisting in the production of diagrams and in the checking of calculations.

In the Appendix, Tables A and D are reproduced from D. V. Lindley and J. C. P. Miller (1966), *Cambridge Elementary Statistical Tables*, published by Cambridge University Press. Table B is from H. R. Neave (1978), *Statistical Tables for Mathematicians, Engineers, Economists and the Behavioural and Management Sciences*, published by George Allen & Unwin, Ltd. I am grateful to the Literary Executor of the late Sir Ronald A. Fisher, F.R.S., to Dr. Frank Yates, F.R.S., and to Longmans Group Ltd., London, for permission to reprint Table C from their book *Statistical Tables for Biological, Agricultural and Medical Research* (6th edition, 1974). Table E is taken from P. J. Taylor (1977), *Quantitative Methods in Geography: An Introduction to Spatial Analysis*, published by Houghton Mifflin Company, and is based on original tables in the *Psychological Bulletin* and the *Annals of Mathematical Statistics*. The material in Table E is reprinted by permission of the American Psychological Association and the Institute of Mathematical Statistics. Tables F to J are based on material tabulated in J. H. Zar (1974), *Biostatistical Analysis*, published by Prentice-Hall Inc. The original sources of Tables F to J are mostly in the *Journal of the American Statistical Association* and are reproduced by permission of the American Statistical Association.

Acknowledgements

Contents

1

Introduction. Quantification in a Context

QUANTIFICATION and statistics are present, in some form or other, in all branches of Geography. It follows, therefore, that no geographer can be adequately prepared to pursue the subject unless he or she has at least an elementary knowledge of this group of techniques. A knowledge of quantitative and statistical techniques is necessary to obtain full benefit from the literature of Geography, which increasingly (and quite rightly) tends to expect from the reader a basic level of technical competence. These techniques are also necessary if one wishes to be a practitioner of Geography, no matter at what level this may be attempted. The manual proceeds on the assumption that the best way to understand a technique is to use it, and that the best way to appreciate the advantages and limitations of its use in Geography is to apply it to geographical problems. First, however, this chapter provides a framework within which the range of techniques can be viewed and their interrelationships clarified.

Why should Geography be quantitative?

Geography employs quantitative techniques for a very good reason. Nothing is wrong with a qualitative statement, but it will carry more weight if it is possible to make a statement quantitatively; that is, in a mathematical language rather than in words. Why is this so?

Ideally, the aim of a geographical statement is to convey unbiased, objective information. The advantage of a quantitative statement is its *precision*, which allows less room for subjective bias to enter into the construction and interpretation of the statement. Consequently, a quantitative statement is more amenable to verification, more easily compared with other statements, and generally more suitable for testing hypotheses and developing theory by scientific method. In these respects geographical statements are no different from those of any other science.

Superiority of a quantitative statement over a qualitative one cannot be taken for granted however. Quantitative techniques, like all others, can be misused. A quantitative statement is superior only if the following two conditions, namely *validity* and *accuracy*, are met. For any statement to be valid, it must express the true meaning of what it is attempting to represent. For example, if the aim is to compare the 'standard of living' in different countries throughout the world, then one quantitative measure that is available for most countries is the 'income *per capita*' of the population. The question is whether or not 'income *per capita*' is a valid measure of 'standard of living'. In the minds of many, 'standard of living' has very little to do with income in monetary terms. Furthermore, the imprecision of the qualitative phrase 'standard of living' may be necessary to adequately describe this complex concept. Thus the availability of data, the ease of measurement, the simplicity of many quantitative measures, and related features of quantification, should not be confused with the more important notion of validity.

Just as a precise, quantitative statement may not be valid, so it may not be accurate. Precision (exactness) does not necessarily ensure accuracy (correctness), although lack of precision precludes the highest accuracy and, other things being equal, a precise statement is more likely to be accurate than an imprecise one. A simple example of an inaccurate but precise statement is one based on measurements with a faulty instrument, but there are many other potential sources of error. The results of shooting at a target provide good analogies of accuracy and precision and illustrate the differences between qualitative and quantitative statements in terms of these two concepts (Fig. 1). Precision is represented by the degree of scatter of the individual shots; accuracy is the analogue of closeness to the bull's eye. Figure 1(A) is as precise as Fig. 1(B) but is not as accurate; Fig. 1(C) is not as accurate or as precise as either 1(A) or 1(B); Fig. 1(D) has a wide scatter, but is centred on the target, and is therefore not very precise but quite accurate. Figures 1(B) and 1(D) have much in common with good quantitative and good qualitative statements, respectively. Geographical statements, which are equivalent to the graphical representations in Fig. 1, may be listed:

(a) 9 million km² in northern Siberia has a tundra vegetation cover;
(b) 3 million km² in northern Siberia has a tundra vegetation cover;
(c) most of northern Siberia is covered with Boreal coniferous forest (taiga);
(d) most of northern Siberia is covered with tundra vegetation.

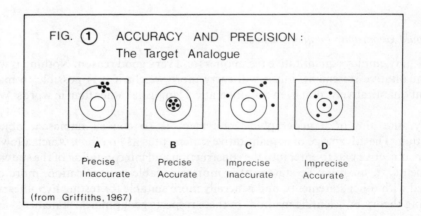

FIG. ① ACCURACY AND PRECISION : The Target Analogue

A	B	C	D
Precise	Precise	Imprecise	Imprecise
Inaccurate	Accurate	Inaccurate	Accurate

(from Griffiths, 1967)

According to Walter (1973), (b) is accurate and precise and therefore the best of the four statements. Both (a) and (c) are seriously in error, the former being inaccurate though precise, the latter being inaccurate and imprecise. Statement (d) is accurate but its imprecision allows a range of possible interpretations and could be a source of confusion.

The greater role of scientific method

Quantification is rarely an end in itself but is an integral part of the method by which reliable knowledge is accumulated. Any science proceeds by a cycle of *observation* and *hypothesis** from which emerges scientific order, or *explanation*, in which observed facts are given meaning within a conceptual framework (Fig. 2). Observations are not made without the benefit of experience or existing hypotheses, and hypotheses are not derived entirely in isolation from the real world. Moreover, our hypotheses are always subject to improvement. Thus scientific method should not be viewed as having a starting-point or an end-point, but as a progressive process of successive approximation.

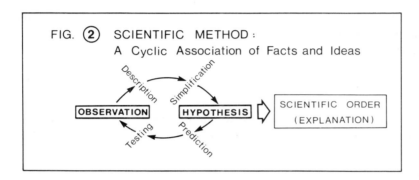

FIG. ② SCIENTIFIC METHOD :
A Cyclic Association of Facts and Ideas

Description, quantitative or qualitative, may involve and be followed by some kind of *simplification* and the *formulation of hypotheses*. Thus we might classify the phenomena and put forward the characteristic features of a small number of groups rather than describe a large number of individuals in turn. In this way, hypotheses with a strong *inductive* component could be suggested from observation of the real world. Alternatively, hypotheses may have a strong *deductive* component, being derived in large part from the existing body of reliable knowledge (rather than from observation directly).

The expected consequences of hypotheses—*predictions*—can be made about similar phenomena, in the same place, or in different places, or at different times (in the past or future). Hypotheses, no matter how derived, are tested by comparing such predictions against *independent evidence*; for example, new observations which were not used in the original formulation. Failure to stand such a test—*refutation*—calls for a new or at least a modified hypothesis, to take account of the independent evidence. On the other hand, well-tested hypotheses may be raised in status and may be said to provide general explanations or principles, which form part of the body of reliable knowledge.

Hypotheses (together with descriptions, predictions and other aspects of scientific method) are ideally *simple*, *quantitative* and *general*. Unfortunately, a hypothesis does not always possess all three properties. For example, simplicity often results in the recognition

---

* A hypothesis is a general proposition about all the individuals of a particular sort; a testable conjecture; a potential solution to a problem; a model.

of many exceptional individuals and it may be difficult to quantify the necessary range of variation to achieve generality. Although words can be used for logical reasoning and may result in hypotheses and the establishment of scientific order, words are clumsy tools. Numbers and mathematical signs are less obstructive to precise and accurate reasoning, as outlined above.

Scientific method is not the only basis for understanding, but it provides us with the 'most consistent, coherent and empirically justified body of information upon which to base such understanding' (Harvey, 1969). This is how scientific progress is made; geographers should be satisfied with nothing less. Geography is no less dependent than any other science on the accumulation of reliable knowledge. However, its subject matter does differ in some important respects from that of 'conventional' or 'hard' sciences, such as Physics and Chemistry. The complexities of geographical phenomena in space and time (particularly on the social science side of the subject) are often very difficult to tackle quantitatively, and partially account for the slower progress of Geography relative to some other sciences.

The component quantitative techniques

Four broad categories of quantitative techniques are considered in the manual and will be outlined here: *measurement, sampling, descriptive statistics* and *inferential statistics*.

(i) MEASUREMENT

Measurement is the quantitative description of single objects and in Geography involves such diverse topics as the use of instruments and the construction of questionnaires. A particularly important aspect is the *level* of measurement (or order of precision) that is attempted. Measurement is most commonly conceived as *interval scale* measurement; that is, with an exact difference between phenomena measured on that scale. For example, the three highest peaks, measured on an interval scale are:

Everest	8840 m (29,002 feet)
Mount Godwin-Austen (K2)	8611 m (28,250 feet)
Kangchenjunga	8579 m (28,146 feet)

In Geography, much use is made of two other levels of measurement: *ordinal scales*, on which objects are measured according to their rank-order (that is, they are placed in increasing or decreasing order); and *nominal scales*, which indicate differences in kind rather than degree. Ordinal scale measurement of the three highest peaks results in the statement that Everest is higher than Mount Godwin-Austen, which is higher than Kangchenjunga. Note that this statement does not say by how much one peak differs from another. On a nominal scale, all three peaks might simply be described as belonging to the same class of phenomena—'mountains'—without even an implied ordering. Ordinal and nominal scales are particularly widely used in Human Geography. For example, the description of people as belonging to 'upper', 'middle' or 'lower' classes, or countries as being 'developed' or 'developing', involves the use of ordinal scales. Examples of nominal scales include land-use categories, industrial types and personal occupations.

Just as quantitative statements are superior to qualitative statements (because of higher precision), so interval scale measurement is superior to ordinal scale measurement, which is in turn superior to nominal scale measurement, for the same reason. Quantitative data resulting from different levels of measurement often require different techniques for further analysis because many techniques are specifically designed to handle data at a particular level of measurement. However, whenever possible, measurements should be made on an interval scale as more precise information about the objects is involved in the measurement. For this reason most emphasis is given in the manual to techniques requiring interval scale data.

(ii) SAMPLING

Sampling is concerned with the choice of an object or set of objects for measurement. Geographers have always utilized sampling although they have not always sampled quantitatively or objectively. Case studies (such as a detailed examination of one farm within an agricultural region, a detailed study of one section exposed in a particular landform, or a detailed study of the development and economic consequences of a single hurricane) are a traditional approach in all branches of the subject. Case studies are samples of one object (sample size = 1); more usually, sampling is concerned with the choice of a set of objects of the same sort, because it can be dangerous to rely on results derived from a single example.

The aim of sampling is usually to obtain an unbiased or *representative sample* of the larger *population* of objects from which the sample is drawn, thereby ensuring that the sample has some general significance. It is often grossly inefficient and unnecessarily time-consuming to measure the whole population, because it is possible to arrive at a sufficiently accurate and precise estimate from a sample. It is essential, however, that a sample is representative and not subject to bias, such as the personal prejudices of the investigator.

Sometimes it is virtually impossible to measure the whole population—such as all the pebbles on a beach or all the people in a city—even if it were desirable to do so. Conservation can also be an additional incentive to sample as efficiently as possible (particularly in the study of soils and vegetation, where the object of study is sometimes destroyed, or at least disturbed, during sampling and measurement). The primary consideration in sampling must remain the representativeness of the sample, which determines to a great extent the value of results following from the analysis of data. All subsequent quantitative techniques that are used to analyse data in this manual assume the data are representative, unbiased samples.

(iii) DESCRIPTIVE STATISTICS

Descriptive statistics are *quantitative summaries* of the measurements made on a set of objects. Examples include totals, averages, percentages and measures of the variability of a set of objects. Assuming that adequate measurement and suitable sampling techniques have been employed, then the result should be a lot of reliable quantitative data. Such information, perhaps based on 500 questionnaires collected from a survey by interview, or 1000 measurements of pebble size from a beach, is of little use in the form of raw data but is more easily assimilated and communicated to others in a summarized form. Moreover, once summarized, one data set is more easily compared to other data sets.

By the judicious substitution of a few descriptive statistics for many individual measurements generalizations are made but, inevitably, some information is lost. It is the information that is lost and the information that is retained which determines the advantages and limitations of the various descriptive statistics. The manual begins with an introduction to descriptive statistics (see Chapter 2).

(iv) INFERENTIAL STATISTICS (PROBABILISTIC STATISTICS)

The geographer's need of inferential statistics is a result of his need to take account of the variability of individual objects within samples, which reflects, in a representative sample, the variability of the underlying population. We require to know the confidence that we can place in statements about populations derived from samples. Inferential statistics meet this requirement by making such statements with reference to a standard—the likelihood of a given sample outcome having occurred by chance.

The likelihood of occurrence by chance is determined from sample descriptive statistics in conjunction with the laws of probability, and is expressed in terms of *probabilities*. The following three statements are quantitative and probabilistic, mean exactly the same thing, and are given as examples of a simple conclusion from the application of inferential statistical techniques to a sample of pebbles from a beach:

(a) there is a 95 % ($p = 0.95$) probability that any pebble on this beach is greater than 2.0 cm long;
(b) there is a 19:1 (95:5) chance that any pebble on this beach exceeds 2.0 cm in length;
(c) 95 pebbles out of any 100 (950 out of any 1000) on this beach are likely to be longer than 2.0 cm.

The uses of inferential statistics are most commonly described in two closely related ways. These are:

(a) *estimating the properties of populations* on the basis of samples;
(b) *testing hypotheses* about one or more populations from samples.

The three statements given above are examples of the first use. They give an estimate of some property of the population of pebbles and a measure of the confidence that can be placed in it. Use of inferential statistics in the testing of hypotheses might involve the comparison of the pebble sizes on two beaches, based on a sample of pebbles from each. Possible conclusions would be:

(a) there is a $> 95\%$ ($p > 0.95$) probability that the two beaches differ in terms of their pebble sizes;
(b) there is a $> 19:1$ ($> 95:5$) chance that pebble sizes differ between the two beaches;
(c) there is a $< 5\%$ chance of there being no difference between the two beaches in terms of their pebble sizes.

Inferential statistics provide, therefore, a means of measuring the uncertainty associated with sampling from populations, and a precise measure of the confidence that can be placed in results based on representative samples.

Exercise 1. Problems of measurement in the invention of shape indices for landforms.

Background

Shape or form is often a more interesting and more important aspect of phenomena than mere size. The terms 'geomorphology' (earth-form study) and 'landform' emphasize this importance. Shape is difficult to quantify, however. The same shape may occur at widely different scales. For example, running water naturally tends to meander whether one considers a small streamlet on a waste-tip or the Mississippi river before it enters the sea. The meander landform has a characteristic shape, irrespective of its scale. Any quantitative measure designed to describe the shape of a meander must not be influenced by the size of the feature, but should capture the essentials of 'meanderingness' or 'sinuosity'.

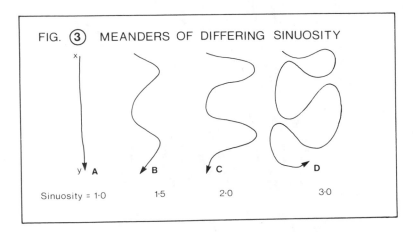

FIG. ③   MEANDERS OF DIFFERING SINUOSITY

A suitable measure or index of sinuosity is illustrated in Fig. 3. It is easy to appreciate that sinuosity increases from A to D and that the straight channel A has minimum sinuosity. In D the length of the channel is great in relation to the straight-line distance between x and y (defining the reach of the river that is being described). It is a small step from this observation to the definition of an index of sinuosity as the *ratio* of the actual length of the channel to the straight-line distance (that is, A/S). The advantage of the index is that it enables one to state precisely the degree of sinuosity of any channel. In Fig. 3, for example, A has a sinuosity of 1.0, C is twice as different from a straight channel than is B, and D is more different from C than is B.

Although shape is independent of size, measurements of size (measurements of length in the example above) are necessary to determine a measure of shape. It is the use of the measures of size as a ratio that removes the effect of size. The following exercise on the invention of shape indices for various landforms requires, therefore, the construction of appropriate ratios.

Practical work

1. Drumlins are ice-moulded landforms composed of glacial till and have been described qualitatively as being shaped like birds' eggs, cigars, airships and upturned spoons. A plan view of part of a drumlin field in Wisconsin, U.S.A., is shown in Fig. 4. Many research workers have used the length:width ratio (L/W) as a measure of drumlin shape. This shape index varies from 14.5 to 1.6 when applied to the individual drumlins in Fig. 4.

FIG. ④ DRUMLINS IN WISCONSIN
(Horicon Quadrangle ,U.S.A.)

(a) What aspect of shape changes as this index increases in value?
(b) Draw drumlins with shape indices of 2.0, 3.0 and 6.0, respectively.
(c) Suggest a name for this index.
(d) Values of this index in two other drumlin fields are:

Region	Maximum value	Minimum value
Central Finland	50.0	2.0
S.W. Scotland	6.0	1.0

Suggest some geomorphological reasons why the shape of drumlins varies between these three areas of Wisconsin, Finland and Scotland.

(e) Name at least one aspect of drumlin shape that is *not* taken into account by the length:width ratio.

FIG.⑤ DIVERSE LANDFORMS

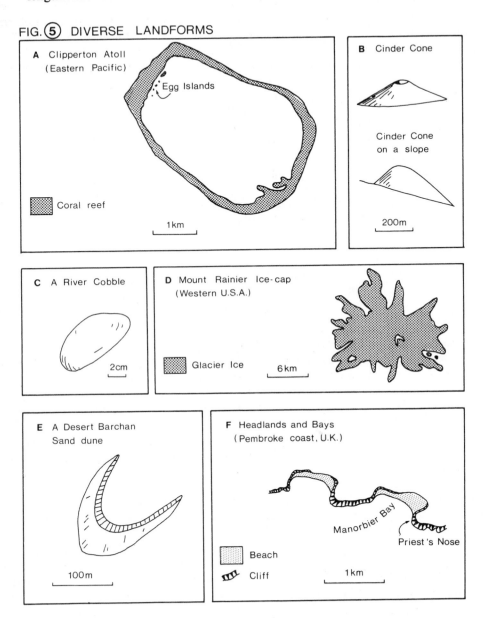

A Clipperton Atoll
   (Eastern Pacific)

Egg Islands

Coral reef

1km

B Cinder Cone

Cinder Cone
on a slope

200m

C A River Cobble

2cm

D Mount Rainier Ice-cap
   (Western U.S.A.)

Glacier Ice   6 km

E A Desert Barchan
   Sand dune

100m

F Headlands and Bays
   (Pembroke coast, U.K.)

Manorbier Bay

Priest's Nose

Beach

Cliff   1 km

2. Drainage basin shape is commonly measured by the length:width ratio. Two other widely used indices are:

$$\frac{I}{C} = \frac{\text{Area of the largest circle that can be drawn within the watershed}}{\text{Area of the smallest circle that can be drawn around the watershed}}$$

and $\dfrac{L}{A} = \dfrac{\text{Length of the basin}}{\text{Diameter of a circle with the same area as the basin}}$.

(a) Discuss the differences between the indices in the context of drainage basin shape. (It would be advisable to draw some 'model' shapes to assess the application of the three indices.)

(b) Considering each index in turn, what differences in index value would be expected for basins in the following regions:

    (i) A region of uniform sedimentary rock and a maritime temperate climate.

    (ii) A region recently deglaciated after a period of intense glacier erosion.

3. Theoretically, if a landform can be described qualitatively it should also be possible to describe it quantitatively. Figure 5 gives a selection of geomorphological phenomena, some of which have characteristic shapes, but few of which have been described by suitable shape indices. Make notes on possible indices for these features, paying particular attention to the following points:

(a) Precise definitions of what must be measured.

(b) The aspect of shape that each index is describing.

(c) Any limitations of your indices.

Exercise 2. Scientific method in the analysis of the Roman road network in England and Wales.

Background

Historical and archaeological research suggests that London was probably the economic centre of Roman Britain, an importance attributed in part to its geographical position and the excellence of its communications. Provincial administration was centred on a large number of towns of supposed lesser economic importance than London: the 'colonia' (Colchester, Gloucester, Lincoln and York) and the 'municipium' (St. Albans), which were communities of Roman citizens; and fifteen 'civitates' or provincial capitals. Major legionary forts were located at York, Chester and Caerleon. The road network of Roman Britain (Fig. 6A) was a carefully planned system, linking these civil and military centres of occupation.

This exercise applies a simple quantitative technique to the Roman road network, as a means to improving our understanding of London in relation to the rest of Britain at that time. Particular attention is paid to the care required in the formulation of hypotheses and to the relationship between hypotheses, independent evidence and assumptions.

In Fig. 6B the original network has been converted into a 'polar network', that is a network centred on a particular point, in this case London. This has been achieved by identifying 'indifference' points that are equidistant from London by alternative routes. At indifference points the original network has been cut. In Fig. 6C the polar network has been 'ordered', as indicated in the inset to Fig. 6B. In this way the routes centred on London have been quantified using an ordinal scale level of measurement.

Practical work

1. Is the converging pattern of routes in Fig. 6C evidence for the focal position of London in the Roman road network?

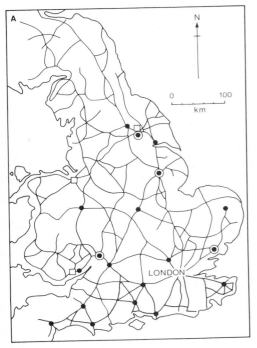

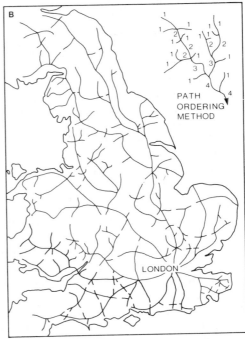

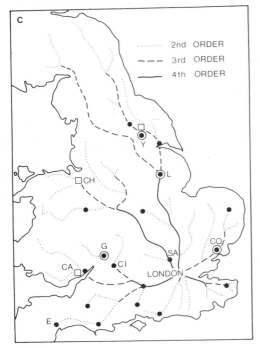

A THE ROMAN ROADS
B THE POLAR NETWORK:
Centred on London
C THE ORDERED NETWORK
(All from Dicks . 1972)

⊙ Colonia
● Civitate
□ Legionary Fort

CA	CAERLEON	G	GLOUCESTER
CH	CHESTER	L	LINCOLN
CI	CIRENCESTER	SA	ST. ALBANS
CO	COLCHESTER	Y	YORK
E	EXETER		

FIG. ⑥ NETWORK ANALYSIS OF
BRITISH ROMAN ROADS

2. What meaning can be attached to the order of a route in Fig. 6C, and, for the ordering system to be meaningful, what simplifying assumptions have to be made?

3. Which of the following were involved in the construction of Fig. 6C:

(a) A test of a hypothesis?

(b) The formulation of a hypothesis?

(c) The predicted consequences of a hypothesis?

(d) The description of observations?

(e) An arrangement of data to fit preconceived ideas?

Fully justify your answer in each case, including an explicit statement of any hypotheses that are discussed.

4. The municipium, the four coloniae and seven of the civitates are located on third- or fourth-order routes; all the remaining civitates are on second-order routes. Does the independent evidence of the location of the Roman towns, when viewed in relation to the polar network (Fig. 6C), corroborate the hypothesis that the Roman urban hierarchy was dependent on London? Fully explain your answer.

5. The following three routes are thought to have been the main arteries of Roman Britain:

  (i) The northern route (Ermine Street), linking London, Lincoln and York.

 (ii) The midland route (Watling Street), linking London, St. Albans and Chester.

(iii) The western route, linking London to Silchester with branches to Gloucester and Caerleon.

Does the high order of these routes in Fig. 6C support any hypothesis? If so, what is the hypothesis, and why is it supported? If not, why not?

6. If you were given polar networks based on the same roads, but centred on different foci, what additional information might be forthcoming?

7. Figure 6C bears some similarity to Britain's contemporary motorway system. Would it be valid to suggest that the same terminals and the same problems of determining the most practicable routes to London therefore governed the engineers' choices in both ages?

8. Would the discovery of new Roman roads influence the results obtained?

9. Discuss the bearing of the following points on the validity of possible conclusions from Fig. 6:

(a) At stage 2 of the analysis (Fig. 6B) two routes of major importance in Roman Britain were cut, namely:

   (i) The straight-line route linking Exeter to Cirencester and Lincoln (the Fosse Way).

  (ii) The east-west route linking Colchester, St. Albans and Cirencester.

(b) There were important links between Roman London and continental Europe, which are not reflected in the ordering of routes in south-east England.

(c) York, whose legions were strategic reserves for the defence of the north, is entirely unconnected with the western half of Hadrian's Wall in Fig. 6C.

10. The analysis in Fig. 6 was originally carried out by Dicks (1972) who concluded:

Decomposing the complex network of the Roman road system of Britain into a polar network centred on London, corroborates much of our knowledge of Roman Britain, emphasizes the primacy of London, and suggests the value of network analysis as an exploratory tool in historical geography.

Give your views on the validity of these conclusions.

2
Measures of
Central Tendency

GIVEN a set of measurements derived from a set of objects of similar sort, what descriptive statistics are necessary to describe adequately the data? These can be appreciated by reference to the *frequency histogram* (Fig. 7) which is a graphical representation of the data. The figure indicates how many of the measurements (vertical axis) fall within certain limits on the scale of measurement (horizontal axis). The histogram has a form or *distribution* that is commonly found with data derived from objects of interest to geographers; there is a peak which indicates the most frequently occurring measurement and the distribution tapers off with relatively few extremely high or extremely low values. Sometimes the distribution is asymmetrical, but for the present a symmetrical distribution will be assumed. The distribution in Fig. 7 is centred on a particular point on the scale of measurement and has a certain amount of spread (variability) along that scale. Two summarizing measures would adequately describe these aspects of the distribution: first, a measure of *central tendency*; second, a measure of variability or *dispersion*.

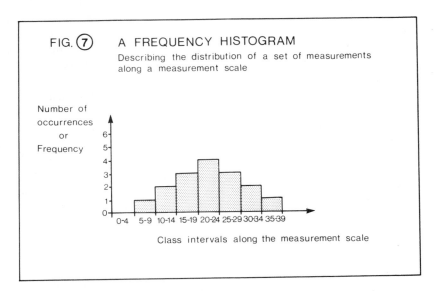

FIG. (7)   A FREQUENCY HISTOGRAM
Describing the distribution of a set of measurements along a measurement scale

Number of occurrences or Frequency

Class intervals along the measurement scale

The most useful measure of central tendency is the *mean*, commonly termed the average. To calculate the mean of a set of measurements, the individual measurements are added together and then divided by the number of measurements. In symbols, this calculation becomes:

$$\bar{x} = \frac{\Sigma x}{n}$$

where, x = an individual measurement,
  Σx = the sum (total) of the individual measurements,
  n = the number of measurements (the sample size),
  $\bar{x}$ = the mean (average) of the sample.

The mean of the whole population, from which the sample was taken, should be represented, as μ, and the number of individual measurements in the population by N. These population values are not usually known, however.

Two other measures of central tendency, generally less useful but having important applications associated with particular types of distribution and with ordinal and nominal scales of measurement, are the *median* and the *mode*. The median is the middle value when the individual measurements in a set of data are arranged in rank order (that is, in order of increasing or decreasing value). The mode is simply the most commonly occurring value in the set (or the highest column in the histogram). If the data form a symmetrical distribution (as in Fig. 7) the three measures of central tendency will be approximately the same.

Mean annual rainfall is a very widely used measure of central tendency, which describes the amount of rainfall that has occurred, on average, over a period of years. This value usually gives a good indication of rainfall conditions, but under some conditions this mean can be misleading. For example, at Iquique, a meteorological station in the Atacama Desert (northern Chile), 63.5 mm (2.5 inches) of rain fell in a few hours on 22nd June 1911, in a series of otherwise rainless years. This amount of rain falling on one occasion in 8 years gives a mean annual rainfall of almost 8.0 mm. But in seven out of the eight years no rainfall was experienced! The median (the middle value) and the mode (the most common value) are both 0.0 mm in this example and are surely more representative of desert conditions that the mean. More generally, it can be said that the mean is the most sensitive of the three measures of central tendency to *extreme cases*. One more year with, say, 30 mm of rain would not alter the median or the mode at the desert station, but the mean would be greatly changed.

The mean also reacts differently to asymmetry of the distribution. Figure 8 shows two basic ways in which distributions may depart from symmetry. A positively *skewed distribution* (Fig. 8A) has more measurements concentrated at the low end of the measurement scale; a negatively skewed distribution (Fig. 8B) has more cases at the high end of the scale. In both examples the mean is closest to the centre of the range of the measurements (the range being the part of the measurement scale that is occupied by the measurements) and farthest from the 'peak' of the distribution. The mode is furthest from the mean and is, by definition, the highest column in each histogram. The median is intermediate in its sensitivity to skewness.

One situation in which the mean and median will both give a totally misleading impression is when a distribution is bi- or *multi-modal*. An example is given in Fig. 9, which shows a bimodal distribution (two modes). In this example the mean and the median not only give no indication of the two peaks but suggest a central tendency at a

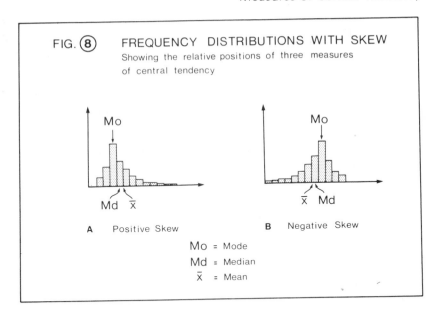

FIG. (8)   FREQUENCY DISTRIBUTIONS WITH SKEW
Showing the relative positions of three measures
of central tendency

A   Positive Skew   B   Negative Skew

Mo = Mode
Md = Median
$\bar{X}$ = Mean

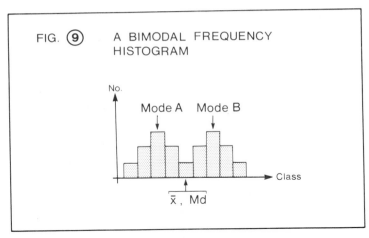

FIG. (9)   A BIMODAL FREQUENCY
HISTOGRAM

point on the measurement scale where very few individual measurements are found. If such a distribution were to be found in reality, then each mode would have to be noted (as will be necessary in Exercise 3). In general, it is best to examine the distribution of individual measurements, by plotting a histogram, before any measure of central tendency is calculated. Other things being equal, however, the mean is to be preferred as it incorporates more information about the individual measurements (interval scale data).

These niceties are particularly relevant in Geography because geographical data are often far from symmetrical, there are often extreme cases, and the measurements are often only available on ordinal or nominal scales. In short, geographical data are statistically 'dirty', as opposed to the 'clean' data sets often available in the 'hard' sciences. This makes for difficulties when we wish to compare two or more 'dirty' data sets. An actual example where such effects as have been discussed above could be crucial is provided in Fig. 10.

The histograms describe maximum valley-side slope angles in badland topography at Perth Amboy, New Jersey, in 1949 and in 1952, respectively. After 4 years of erosion mean slope angle appears to have been reduced by 0.2°. It may well be, however, that this difference is the result of extreme cases or a difference in symmetry of the distributions.

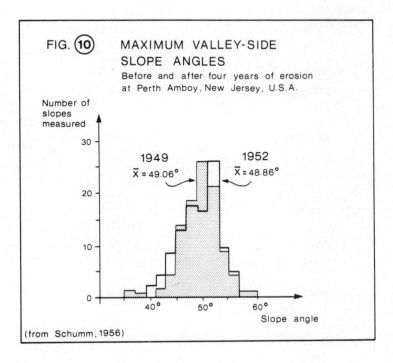

FIG. ⑩   MAXIMUM VALLEY-SIDE SLOPE ANGLES

Before and after four years of erosion at Perth Amboy, New Jersey, U.S.A.

Number of slopes measured

1949 $\bar{X} = 49.06°$ 1952 $\bar{X} = 48.86°$

Slope angle

(from Schumm, 1956)

Exercise 3. Application of histograms and use of the mode in the reconstruction and dating of glacier fluctuations in Swedish Lappland.

Background

In the Kebnekaise Mountains of Swedish Lappland, twenty-three small glaciers were investigated and up to eight crescentic end-moraines were found to have been deposited within a few kilometres of each glacier snout. The moraines are present-day field evidence of past variations in size of the glaciers, for each time that a glacier increases in size it tends to deposit a morainic ridge at the limit of the advance. By dating the end-moraines, it is possible to reconstruct the recent history of glacier fluctuations and hence to make inferences about changes in climate over the same period. The moraines can be dated by lichenometry—that is, by the size of the largest lichens growing on them—for the longer a moraine has been deposited, the longer the period of time that has been available for lichen growth, and the larger the diameters of the almost circular lichens tend to be.

An example of the setting of the moraines in front of one of the glaciers is shown in Fig. 11. This glacier has five end-moraines, characterized by lichens up to 22 mm, 28 mm, 40 mm, 88 mm and 160 mm in diameter, respectively. These data indicate that this glacier was larger at least five times in the past than it is today. The most likely cause of a change in the volume of the glacier is a change in climate; in particular, a change in the balance between the accumulation of snow-fall in winter and the ablation or melting in summer.

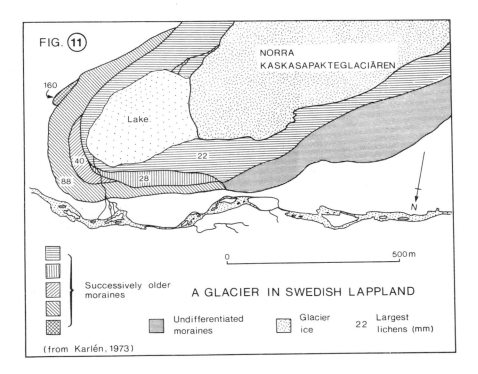

FIG. (11)

NORRA KASKASAPAKTEGLACIÄREN

160

Lake

40

88

22

28

N

0   500 m

Successively older moraines

A GLACIER IN SWEDISH LAPPLAND

Undifferentiated moraines

Glacier ice

22 Largest lichens (mm)

(from Karlén, 1973)

Figure 12 shows the relationship between lichen size and moraine age, established by examination of the size of lichens on surfaces of known age in the same region (young moraines that were observed being formed, mine spoil-heaps, buildings and railway workings of known age).

By use of a histogram to summarize graphically the lichen-size data from all twenty-three glaciers, it is possible to elucidate much concerning the recent glacier and climatic fluctuations. In particular, a mode indicates that a large number of glaciers advanced to form a moraine at about the same time and is therefore evidence for fluctuations of regional, rather than local, significance.

Practical work

1. The data given in Table 1 are the largest specimens of *Rhizocarpon geographicum* (the map lichen) growing on ninety-three moraines in front of twenty-three glaciers in Lappland.

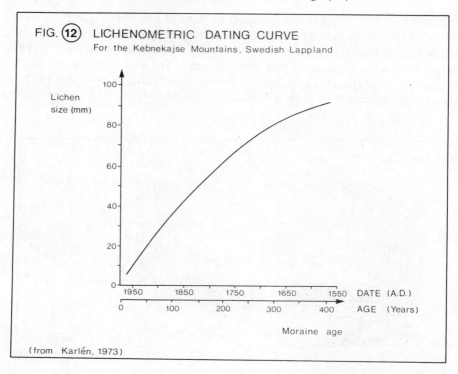

FIG. ⑫ LICHENOMETRIC DATING CURVE
For the Kebnekajse Mountains, Swedish Lappland

Lichen
size (mm)

Moraine age

(from Karlén, 1973)

TABLE 1. *Lichen sizes on moraines in front of glaciers in northern Sweden*

Glacier	Largest lichen (mm) on each moraine
1. Kårsajökeln	21, 42, 58, 88, 155
2. Blaisen	21, 31, 90, 355
3. Rabots glaciär	21
4. Riukojietna	22, 26, 285
5. Storlaciären	21, 30, 60, 75, 85, 185
6. Norra Kaskasapakteglaciären	22, 28, 40, 88, 160
7. Nipalsglaciären	20, 62, 93, 180
8. Östra Pyramidglaciären	19, 38, 75, 176
9. Tjäktjapakteglaciären	21, 85, 178, 320
10. Kuoblavaggeglaciären	14, 100
11. Isfallsglaciären	18, 31, 60, 77, 86, 180
12. Sydöstra Kaskasatjåkkoglaciären	16, 30, 56, 76, 170
13. Östra Kaskevaggeglaciären	13, 62, 72, 87
14. Mellersta Kaskevaggeglaciären	17, 85
15. Enquists glaciär	14, 32, 60
16. Vaktpostglaciären	21, 33, 43, 70, 85, 180, 260, 380
17. Tarfalaglaciären	60, 195, 260
18. Södra Kaskasapakteglaciären	14, 53, 90
19. Björlings glaciär	20, 33, 84, 192, 201, 256, 380, 410
20. Östra Repiglaciären	21, 61, 77, 172
21. Knivglaciären	21, 31, 76, 158
22. Passglaciären	13, 61
23. Kitteldalsglaciären	16, 31, 55

(After Karlén, 1973.)

(a) Draw a histogram to summarize the data. Number of moraines (vertical axis) should be plotted within lichen-size classes (horizontal axis). A 4-mm class interval is recommended. (The class interval should be no larger than this because the data for individual glaciers indicate that distinct moraines differ in lichen size by as little as 4 mm.)

(b) Describe, in some detail, this multi-modal distribution. Consider, in particular:

    (i) The overall distribution.

    (ii) Whether there are broad groups within the general pattern.

    (iii) To what extent minor modes are meaningful.

2. Relate your histogram to the graph in Fig. 12 and summarize the *ages* of the moraines. Your answer should include:

    (i) A description involving precise dates where this is justifiable.

    (ii) Some tentative suggestions about the age of the older moraines.

3. Given that a distinct mode indicates that many glaciers produced moraines (at the limit of an advance) at the same time, what can be concluded about glacier fluctuations from the spacing between modes?

4. What can be concluded about glacier fluctuations from the relative heights of the various modes?

5. Only two glaciers (numbers 16 and 19) have as many as eight moraines; one glacier (Rabots glaciär, number 3) has only one moraine, and this moraine is a very recent feature. State at least two alternative hypotheses to account for these observations.

6. Comment on the limitations of this investigation (at any stage) and suggest methods that might be used in the field to improve our understanding of past glacier and climatic fluctuations in this area.

Exercise 4. Comparison of measures of central tendency in the study of a fan of erratics in the Central Lowlands of Scotland.

Background

About 10 km north-east of Glasgow (near Lennoxtown) there is a small outcrop of a distinctive type of intrusive rock known as essexite. The last ice sheet to move across the area eroded and then deposited fragments of essexite in the form of a fan of erratics eastwards across the Central Lowlands of Scotland. Today, essexite erratics are common in walls throughout the region, having been removed from the fields by farmers, along with other stones above a minimum size. This readily accessible source of erratics of known origin permits the quantitative investigation of changes in the properties of the erratics with distance from the outcrops. These changes may in turn lead to inferences about processes and rates of glacial transport and erosion.

The location of the essexite outcrop and the points at which walls were examined are shown in Fig. 13. Each point is located in the centre of a 200-m section of wall. There are no other outcrops of essexite known to have given rise to erratics in this area, and the outcrops have never been quarried. The majority of the walls were built in the late eighteenth century and are assumed to be a representative sample of glacially transported material.

This exercise is concerned with an investigation of the variation in size of erratics with distance from the outcrops. The study area in Fig. 13 has been sub-divided into zones of

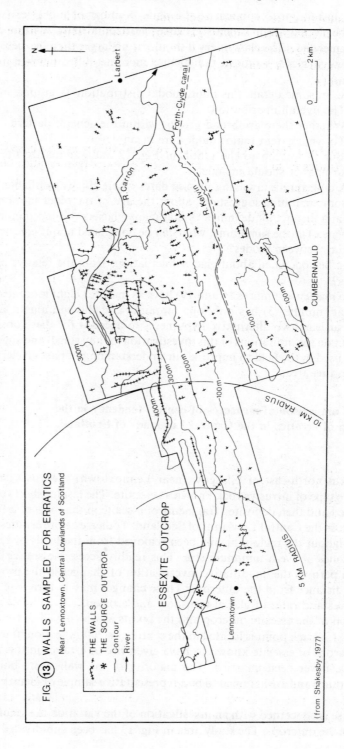

FIG. (13) WALLS SAMPLED FOR ERRATICS

Near Lennoxtown, Central Lowlands of Scotland

✕✕ THE WALLS
✳ THE SOURCE OUTCROP
— Contour
➔ River

(from Shakesby, 1977)

increasing distance from the outcrops; each zone being 2 km wide. Different measures of central tendency are useful for obtaining a generalized and precise measure of the size of erratics with increasing distance from the outcrops. The high variability of erratic size at a given distance makes such measures particularly appropriate for the recognition of any pattern in the data.

Practical work

1. The data given in Table 2 represent the size of individual erratics based on a random sample of 30 erratics from each 2-km zone of the Lennoxtown erratics fan.

TABLE 2. *Size of erratics (cm²) near Lennoxtown, Scotland*

Distance from the outcrops (zone mid-point in km)									
1	3	5	7	9	11	13	15	17	19
1116*	234	900	432	858	195	459	168	216	342
100	728	192	925	322	575	231	153	420	162
130	784	189	403	544	525	464	105	240	182
588	270	315	405	368	330	289	180	100	200
920	200	512	396	360	252	330	779	247	150
561	1404	462	560	630	425	570	768	128	300
500	170	198	416	442	432	528	216	540	130
432	136	285	330	198	578	170	119	594	160
440	351	450	380	680	384	242	322	160	208
112	144	735	108	520	264	392	378	286	252
234	450	736	198	391	392	437	84	216	703
208	261	480	380	896	288	459	165	448	208
70	250	672	910	520	594	1824	144	380	171
306	660	140	437	1440	588	440	350	459	868
169	150	238	432	187	286	375	264	90	200
208	75	325	475	540	56	638	165	156	135
1363	144	551	378	480	70	132	156	135	255
288	527	406	925	308	136	75	247	225	180
322	126	252	209	322	352	437	171	304	91
225	98	289	784	544	234	322	261	144	522
660	77	189	336	434	1080	288	779	152	1880
130	650	644	144	368	56	378	252	143	190
725	144	490	216	198	600	313	714	135	340
532	595	540	290	551	364	294	195	266	66
660	176	483	391	240	162	336	135	286	190
880	200	210	437	486	336	98	289	154	440
450	180	506	486	252	286	756	54	154	198
1440	576	170	380	187	330	150	171	336	242
874	234	437	490	345	513	403	352	378	247
100	513	357	384	459	220	378	442	160	273

* Size of erratics is the area visible in the wall face.

(From R. A. Shakesby, pers. comm.)

Perusal of the data does not reveal any clear pattern; the situation is not much clearer if the whole distribution at each distance is represented graphically in the form of histograms (Fig. 14). Without some additional summarizing of the data, any pattern might remain undetected.

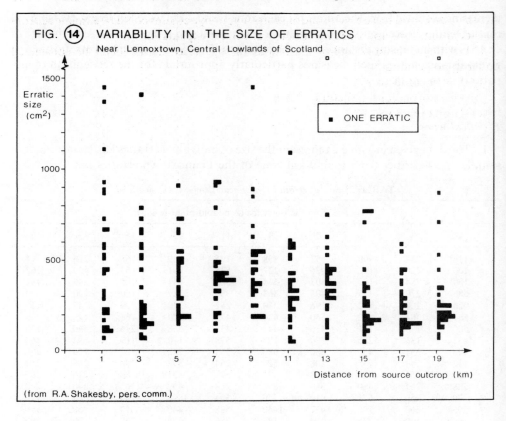

FIG. ⑭   VARIABILITY IN THE SIZE OF ERRATICS
Near Lennoxtown, Central Lowlands of Scotland

Erratic size (cm²)

■ ONE ERRATIC

Distance from source outcrop (km)

(from R.A.Shakesby, pers. comm.)

(a) Calculate the mean size of erratics for each distance, using the above data.
(b) Calculate the median sizes for the same data.
(c) Using Fig. 14, note the modal class at each distance.
(d) Tabulate the measures of central tendency in a manner facilitating comparison.
2. Plot the measures of central tendency on graph paper, with erratic size (vertical axis) against distance from the outcrops (horizontal axis).
(a) Describe the patterns revealed by the three measures of central tendency, paying particular attention to:
    (i) Points of agreement between the measures.
    (ii) Points of disagreement.
(b) Which measure of central tendency shows the most irregular pattern and why should this be so?
(c) Account for the consistent difference between the mean and the median.
(d) Which measure of central tendency is the most suitable in the present context and why?
3. Suggest some possible explanations, in terms of glacial transport or other geomorphological factors, for:
(a) The general pattern observed.
(b) Significant departures from the general pattern.

Are there any good reasons why one or more of your hypotheses could be eliminated from consideration, thus approaching closer to a correct explanation?

4. List any difficulties and/or limitations of this study. These might be considered under three headings:

   (a) Statistical factors.

   (b) Geomorphological factors.

   (c) Human factors.

3

Measures of Dispersion

MOST sets of measurements have a considerable amount of dispersion (also known as variability or spread) which gives any histogram its width. A good quantitive summary of a data set requires, therefore, not only a measure of central tendency but also a complementary measure of dispersion. As well as being an important property of objects of a similar sort, dispersion must also be taken into account when two data sets are being compared. Two sets of measurements may have identical central tendencies but differ greatly in dispersion (Fig. 15). Moreover, a small difference in central tendency between two or more data sets may be of questionable significance if each of the sets possesses a large dispersion with consequent great overlap with other sets (Fig. 10).

Three measures of dispersion will be discussed here, and these may be considered to be complementary to the three measures of central tendency that were considered in Chapter 2:

Measure of central tendency		Measure of dispersion
1.	Mode	Range
2.	Median	Quartile deviation
3.	Mean	Standard deviation

The *range* is the difference between the highest and lowest measurements in a data set. In a histogram, the range is the occupied section of the measurement scale along the horizontal axis (Fig. 7). Like the mode, it is an easily obtained measure containing little information. The biggest limitation of the range is that it is controlled entirely by the extreme measurements at the ends of the distribution, and cannot, therefore, be said to have much generality. The bulk of the measurements in a distribution are usually quite close to the measure of central tendency and a good measure of dispersion should take this fact into account.

The other two measures of dispersion are of much greater importance to the geographer who wishes to make full use of available data. The *quartile deviation* is used in conjunction with the median and, like the median, it is based on ordinal scale measurement (that is, it utilizes rank-order information about individual measurements in a data set). By far the most important measure of dispersion is, however, the *standard deviation*, a measure based on interval scale data (taking the exact value of every individual measurement into account).

24

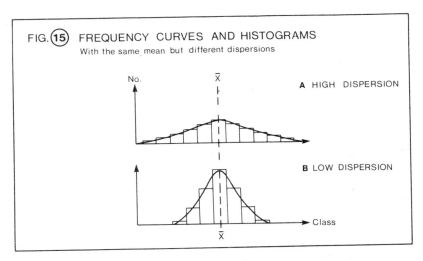

FIG. (15) FREQUENCY CURVES AND HISTOGRAMS
With the same mean but different dispersions

The quartile deviation

The quartile deviation is used to measure dispersion about the median. Just as the median divides a data set into two halves, so the upper and lower *quartiles* subdivide the upper and lower halves of the data. This is illustrated in Fig. 16, which shows individual measurements distributed against a measurement scale in the form of *dispersion diagrams*. It must be stressed that the median and the quartiles are located by counting the number of measurements from the top or bottom. Thus in Fig 16A, the median is the 28th measurement (because there are 55 points on this diagram) and the quartiles are located at the 14th measurements from top and bottom of the measurement scale. If an even number of measurements is involved then the median or quartiles are located half-way between the two middle measurements. The data set, in this case, has been divided into four quarters; it is possible, of course, to make other divisions for particular purposes—quintiles (fifths) or octiles (eighths), for example. In Fig. 16 the individual points are census tracts in the city of Atlanta, Georgia, U.S.A. Each tract has been measured in terms of a quality of life index (derived from information on health, crime, housing, socio-economic status and population density). Separate dispersion diagrams are shown for tracts in which over 50% of the population are white (Fig. 16A) and for tracts in which over 50% of the population are black (Fig. 16B). These diagrams give a clear representation of the difference in quality of life between the two communities, which also are segregated within the city. It is particularly noteworthy that while the ranges of the two communities overlap considerably, there is no overlap between the inter-quartile ranges.

The *inter-quartile range* (the shaded portion in Fig. 16) is the difference, on the measurement scale, between the upper and lower quartiles. Whereas the range encloses 100% of the individual measurements in a data set, the inter-quartile range encloses 50%. The latter, as well as the quartile deviation (which is derived from it), is unaffected by the extreme values that provide the main limitation of the range. The *quartile deviation* is simply half of the inter-quartile range, or:

$$\frac{UQ - LQ}{2} = \frac{\text{Value of the upper quartile minus the value of the lower quartile}}{2}.$$

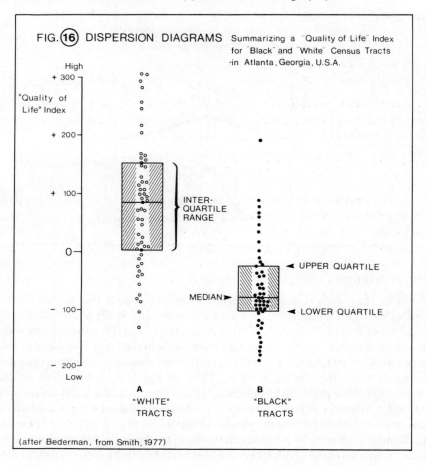

FIG.(16) DISPERSION DIAGRAMS Summarizing a "Quality of Life" Index for "Black" and "White" Census Tracts ·in Atlanta, Georgia, U.S.A.

(after Bederman, from Smith, 1977)

As a general rule, the quartile deviation is preferable to the standard deviation in the same kinds of situation where the median is preferable to the mean, namely:
(a) If the data form an asymmetrical distribution.
(b) If extreme values are a strong influence.
(c) If data are based on an ordinal scale of measurement.

The standard deviation

In Fig. 15 two distributions are shown, with the same mean but different dispersions. They are represented by histograms, and by frequency curves, which should be viewed as smooth curves drawn through histograms. What characteristic of the individual measurements comprising the upper distribution is responsible for its greater dispersion? Therein lies the key to understanding the nature of the standard deviation.

If a lot of the individual measurements are found far from the mean, then the dispersion is great, and it is the amount by which the individual measurements differ from the mean that controls the degree of dispersion. An individual difference or *individual deviation* is represented by:

$$(x - \bar{x})$$

where, x = the value of an individual measurement,
         $\bar{x}$ = the mean of the data set.

The *average deviation*, which is very closely related to the standard deviation, is simply a quantitative summary of the amount by which the whole set of individual measurements differ from the mean:

$$\frac{\Sigma(x - \bar{x})}{n} = \frac{\text{The sum total of all the individual deviations from the mean}}{\text{The number of measurements}}.$$

The average deviation that has just been described is not a suitable measure of dispersion as it stands. When the individual deviations are added up, those measurements that are greater than the value of the mean (positive deviations) would cancel out the measurements less than the mean (negative deviations). In other words, the above formula gives an answer of zero. There are two ways to solve this problem: the first is to ignore the sign (which is not good practice) and deal in deviations irrespective of sign; the second is to take the sign into account mathematically, by squaring the individual deviations, thus converting all positive and negative deviations to positive values. The *standard deviation* takes advantage of this second procedure:

$$s = \sqrt{\frac{\Sigma(x - \bar{x})^2}{n}}$$

where $(x - \bar{x})$ = an individual deviation from the mean,
        $(x - \bar{x})^2$ = the square of an individual deviation,
        $\Sigma(x - \bar{x})^2$ = the sum total of the squared deviations,
        n = the number of individual measurements (the sample size),
        s = the standard deviation of the sample.

The formula for the standard deviation differs in one other respect from the formula for the average deviation. The last step in the calculation of the standard deviation is to take the square root, which is necessary to take account of the fact that the individual deviations have been squared previously. Although this procedure compensates for squaring the individual deviations (by taking the square root) squaring has a disproportionate influence on extreme values. Herein lies a possible disadvantage of the standard deviation, in that a few extreme values can make this measure unusually large.

As a general rule, the standard deviation is best applied when the following conditions are met:

(a) The data have a symmetrical distribution.
(b) Extreme values are not a strong influence.
(c) The data are based on an interval scale of measurement (indeed, the standard deviation cannot be calculated for nominal or ordinal scales).

Although the standard deviation requires more calculation than the quartile deviation, and is not so intuitively interpretable as the average deviation, it is to be preferred in situations where the properties of data are not seriously in conflict with its requirements. It will also be encountered as a vital component of many other techniques in later chapters.

Alternative formulae for the standard deviation

The general formula given above for the standard deviation of a sample is not a good estimate of the population standard deviation when sample size is small. For small samples, therefore, a modified formula is used, which incorporates *Bessel's correction* and gives us the *best estimate of the population standard deviation*. This involves the use of

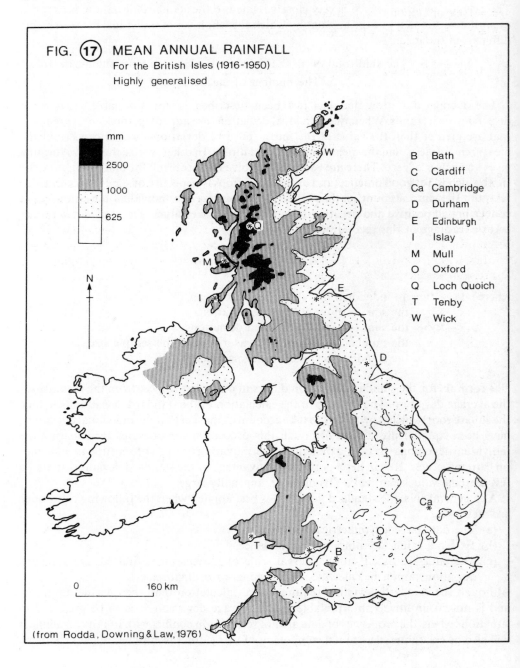

FIG. ⑰ **MEAN ANNUAL RAINFALL**
For the British Isles (1916-1950)
Highly generalised

mm

2500

1000

625

N

B   Bath
C   Cardiff
Ca Cambridge
D   Durham
E   Edinburgh
I   Islay
M   Mull
O   Oxford
Q   Loch Quoich
T   Tenby
W   Wick

0   160 km

(from Rodda, Downing & Law, 1976)

$(n-1)$ rather than (n) in the formula; it makes the resulting measure of dispersion larger and has a greater effect the smaller the sample size.

To distinguish clearly between the various concepts and formulae, the following terms and symbols will be used throughout the manual:

s = the standard deviation of a sample,

σ = the population standard deviation (usually not known),

$\hat{\sigma}$ = the best estimate of the population standard deviation
    (calculated from s by use of Bessel's correction).

The following formulae are appropriate:

$$s = \sqrt{\frac{\Sigma(x-\bar{x})^2}{n}} = \sqrt{\frac{\Sigma x^2}{n} - \bar{x}^2}$$

$$\hat{\sigma} = \sqrt{\frac{\Sigma(x-\bar{x})^2}{n-1}} = \sqrt{\frac{\Sigma x^2}{n-1} - \frac{(\Sigma x)^2}{n(n-1)}}$$

The second version of each formula is quicker on a hand-calculator. The best estimate of the population standard deviation should usually be used as it is based on the principle of 'safety first'; it makes more sense to allow for the possibility that a small sample of measurements may contain less variability than the population from which that sample was drawn. In the worked example given below, the standard deviation of the size of fifteen erratics within 2 km of the source outcrop near Lennoxtown, central Scotland (see Exercise 4), is calculated by all four formulae.

TABLE 3

x	x^2	$(x-\bar{x})$	$(x-\bar{x})^2$
1116	1,245,456	723.6	523,596.96
100	10,000	−292.4	85,497.76
130	16,900	−262.4	68,853.76
588	345,744	195.6	38,259.36
920	846,400	527.6	278,361.76
561	314,721	168.6	28,425.96
500	250,000	107.6	11,577.76
432	186,624	39.6	1568.16
440	193,600	47.6	2265.76
112	12,544	−280.4	78,624.16
234	54,756	−158.4	25,090.56
208	43,264	−184.4	34,003.36
70	4900	−322.4	103,941.76
306	93,636	−86.4	7464.96
169	28,561	−223.4	49,907.56

$\Sigma x = 5886$ $\Sigma x^2 = 3{,}647{,}106$ $\Sigma(x-\bar{x})^2 = 1{,}337{,}439.60$

$$\bar{x} = \frac{5886}{15} = 392.4$$

$$s = \sqrt{\frac{\Sigma(x-\bar{x})^2}{n}} = \sqrt{\frac{1{,}337{,}439.6}{15}} = \sqrt{89{,}162.64} = 298.60$$

$$\sigma = \sqrt{\frac{\Sigma x^2}{n} - \bar{x}^2} = \sqrt{\frac{3,647,106}{15} - (392.4)^2} = \sqrt{243,140.4 - 153,977.76} = 298.60$$

$$\hat{\sigma} = \sqrt{\frac{\Sigma (x - \bar{x})^2}{n-1}} = \sqrt{\frac{1,337,439.6}{14}} = \sqrt{95,531.4} = 309.08$$

$$\hat{\sigma} = \sqrt{\frac{\Sigma x^2}{n-1} - \frac{(\Sigma x)^2}{n(n-1)}} = \sqrt{\frac{3,647,106}{14} - \frac{(5886)^2}{(15)(14)}} = \sqrt{260,507.57 - 164,976.17} = 309.08$$

Exercise 5: Application of dispersion diagrams and measures of dispersion in the study of British rainfall patterns.

Background

Mean monthly and mean annual rainfall data are in common use and are readily understood. The Geography of British rainfall when presented in terms of mean annual values (Fig. 17) presents a striking pattern, is well known, and can be readily explained, particularly with reference to topography. However, mean data can hide great variability, and rainfall is notoriously variable from year to year. It is advisable, therefore, when seeking to describe and explain the climate of a place, to consider measures of dispersion as well as the more usual measure of central tendency.

This exercise seeks to demonstrate the usefulness of dispersion diagrams and measures of dispersion in going beyond mere description, to the inference of processes responsible for the observed patterns. The exercise consists of two parts: first, to clarify the techniques themselves, dispersion diagrams and measures of dispersion are calculated for data from a local meteorological station; second, maps and graphs are presented, which summarize the labours of many such calculations from the remainder of the British Isles.

Practical work

1. The data in Table 4 are monthly rainfall totals for Cardiff (Rhoose airport) for the period 1955–76, a 22-year record.

TABLE 4. *June and December rainfall totals (mm) for Cardiff, South Wales*

	June	December		June	December
1955	135.8	131.3	1966	107.1	119.1
1956	49.8	93.2	1967	20.4	83.5
1957	41.4	41.8	1968	121.6	85.3
1958	85.6	105.7	1969	46.3	67.2
1959	44.2	162.5	1970	71.3	56.4
1960	31.5	131.4	1971	151.3	50.9
1961	40.0	112.4	1972	64.5	97.8
1962	23.4	54.6	1973	51.2	61.1
1963	60.9	53.3	1974	57.0	100.9
1964	55.8	101.6	1975	5.4	38.5
1965	96.3	208.0	1976	32.0	90.0

(From the Meterological Officer, Rhoose airport, pers. comm.)

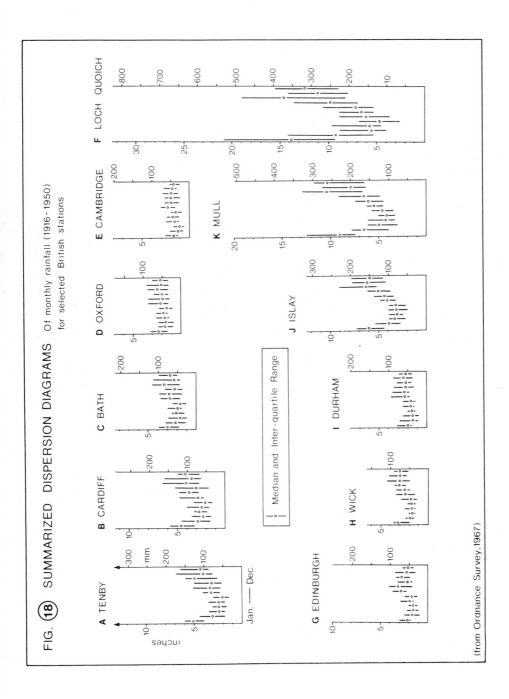

FIG. (18) SUMMARIZED DISPERSION DIAGRAMS Of monthly rainfall (1916-1950) for selected British stations

(from Ordnance Survey, 1967)

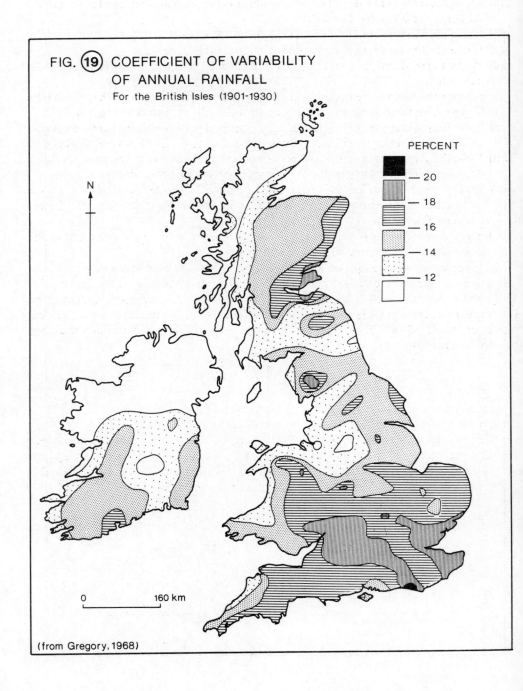

FIG. ⑲ COEFFICIENT OF VARIABILITY
OF ANNUAL RAINFALL
For the British Isles (1901-1930)

PERCENT

— 20

— 18

— 16

— 14

— 12

N

0   160 km

(from Gregory, 1968)

It would be profitable if the reader could obtain local data for this section of the exercise.

(a) Using the above data, draw two comparable dispersion diagrams, one for the month of June, the other for December.

(b) Calculate the median and quartile deviation for each.

(c) Calculate the mean and standard deviation for the same data.

(d) Write a brief description of the climate at Cardiff, incorporating your results from (a) to (c).

2. Examine the summarized dispersion diagrams for selected stations in the British Isles (Fig. 18) and note their location on the map of mean annual rainfall (Fig. 17).

(a) Describe the main features of amount and seasonality of rain along a transect across England and Wales from Tenby, through Cardiff, Bath and Oxford to Cambridge.

(b) Describe and explain the differences between the dispersion diagrams for Loch Quoich, Mull and Islay in the Highlands and Islands of Scotland.

(c) Using the evidence of the dispersion diagrams, consider the validity of the suggestion that eastern stations have a summer maximum of rainfall whereas western stations have a winter maximum.

(d) Suggest some meteorological reasons for differences in the variability and in the seasonality of western and eastern stations.

(e) Is the mean an appropriate measure of central tendency for these data?

3. Figure 19 is a representation of the variability of annual rainfall over the British Isles. The map shows the *coefficient of variability* (or coefficient of variation), a descriptive statistic derived from the standard deviation. The coefficient of variability is the standard deviation expressed as a percentage of the mean; that is, it is a measure of variability, *relative* to the mean:

$$\frac{\text{Standard deviation}}{\text{mean}} \times 100\%$$

(a) Calculate this coefficient for the meteorological station in question 1.

(b) Describe the pattern shown for the British Isles as a whole in Fig. 19.

(c) Explain why the coefficient of variability tends to attain highest values in those areas of the country where the dispersion diagrams indicate that monthly totals vary least from year to year. This part should consider both statistical reasons for this paradox, and a meteorological explanation for the pattern in Fig. 19.

4

Probability Statements and Probability Maps

BECAUSE of the variability of geographical phenomena of a particular sort, descriptive statistics are of only limited value in summarizing geographical data and in the testing of geographical hypotheses. For example, one cannot be 100% sure of the amount of rainfall that will fall at a particular place in a particular year, even though the mean annual rainfall is known precisely from a long period of observations. The mean annual rainfall (and the complementary standard deviation) are the *best estimates* available.

Although it is not possible to be certain of exceeding a given amount of rainfall in any one year (such as next year), it is possible to obtain a precise statement of the *uncertainty* involved. That is, one can be certain of exceeding a given amount at a precisely known *level of probability*. The introduction of probability into Geography requires a consideration of *inferential statistics*. Geographical statements that are probabilistic are clearly superior to deterministic statements because they give additional information about the likelihood of the statement being true or false (assuming that the techniques have been correctly applied and that their data requirements have been met).

In Fig. 7 a data set was represented as a histogram and a frequency distribution curve, both of which were symmetrical. For the remainder of this chapter it will be assumed that data have not only a symmetrical distribution but also a particular type of symmetrical distribution, known as a *normal distribution*. The usefulness of the mean and standard deviation are in large measure a result of their relationship to normal distributions.

If we have a large representative sample from a population that is normally distributed, then there are *fixed probabilities* of any one measurement being within certain areas under the curve (Fig. 20). Intuitively, it is easy to understand that it is more probable that any single measurement will occur near the central tendency of such a distribution than far away from it. It is also clear that all individual measurements are found somewhere beneath the curve. Thus the area under the curve represents 100% probability (or a probability of 1.0). This is shown in Fig. 20A.

Similarly, there is a 50% (or 0.5) probability that any measurement will occur above the mean and you would also expect 50% of the measurements to lie below the mean (Fig. 20B). Because the curve is symmetrical, there is an identical probability of lying below the mean as lying above it. Just as there are 50% of the measurements above and 50% below the mean, so there are fixed proportions of the measurements lying above, below or between any given number of standard deviations from the mean.

Figure 20C illustrates the 'rule of thumb' that approximately 68% of the measurements lie between one standard deviation above the mean and one standard deviation below the

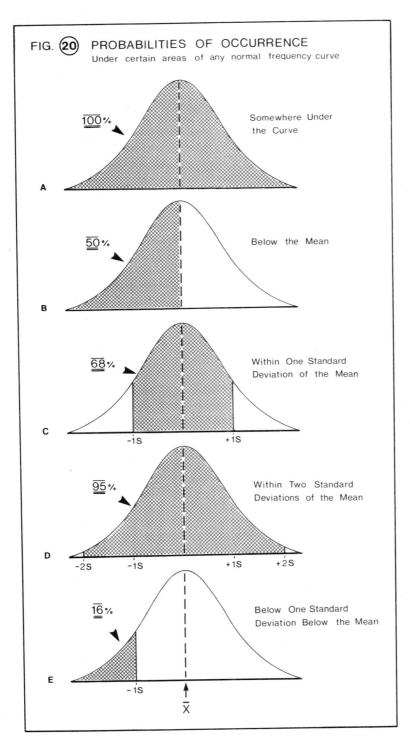

FIG. ㉒ PROBABILITIES OF OCCURRENCE
Under certain areas of any normal frequency curve

$\overline{100}\%$ ▶ Somewhere Under the Curve

A

$\overline{50}\%$ ▶ Below the Mean

B

$\overline{68}\%$ ▶ Within One Standard Deviation of the Mean

−1S   +1S

C

$\overline{95}\%$ ▶ Within Two Standard Deviations of the Mean

−2S   −1S   +1S   +2S

D

$\overline{16}\%$ ▶ Below One Standard Deviation Below the Mean

−1S

$\overline{X}$

E

mean ($\bar{x} \pm 1 s$). To put this another way, there is a 68 % (or 0.68) probability that any one measurement (any one year's annual rainfall, for example) will lie within these limits placed either side of the mean. It is profitable to reflect on the quartile deviation in this respect. The probability that any one measurement will lie between plus and minus one quartile deviation from the median is approximately 50 %; the standard deviation tends to be a more conservative measure of dispersion than the quartile deviation as it encloses a greater proportion of the measurements within the set limits.

Figure 20D shows that approximately 95 % of the measurements lie within plus and minus two standard deviations of the mean ($\bar{x} \pm 2s$). There is a probability of approximately 99.5 % that any one measurement will lie within plus and minus three standard deviations of the mean ($\bar{x} \pm 3s$). It is therefore extremely unlikely that many measurements in any sample will be as far away from the mean as three standard deviations. This fact is sometimes used as a rough check on the calculation of the standard deviation itself (the *three standard deviations check*). If more than one or two measurements lie outside of three standard deviations above or below the mean, then the calculated standard deviation is likely to be in error. Only one measurement in 200 (0.5:99.5) is expected to lie farther than three standard deviations from the mean.

One further point needs emphasis. The areas beneath the curve can be added or subtracted to find the probability of being above, between or below any number of standard deviations above or below the mean. Figure 20E, for example, which can be derived from Fig. 20C, shows that there is an 84 % probability of any individual measurement lying above one standard deviation below the mean (that is, greater than $-1s$), and a 16 % probability of any measurement being below one standard deviation below the mean (that is, less than $-1s$).

So far, only 'rules of thumb' have been discussed. More accurate probability values, which apply to any normal curve and enable the use of probability values associated with fractions of standard deviations, are available in Table A (Appendix). In the table there are two columns of values. These are:

(a) The number of standard deviations from the mean (defined as z).
(b) The probability (p) that any one measurement will lie *above* that number of standard deviations *below* the mean (identical to the probability of lying *below* z standard deviations *above* the mean).

Note that the table always gives the larger area (a probability value greater than 50 %) when a situation like that illustrated in Fig. 20E is being examined. To obtain the smaller area, the tabulated probability is simply subtracted from 100 % (probability 1.0). Thus the probability value opposite $z = 1.0$ in the table is 84.13 % (probability 0.8413), which is the probability that any individual measurement will lie above one standard deviation below the mean (greater than $-1s$).

A worked example using the mean and standard deviation of annual rainfall totals for Edinburgh, Scotland, will serve to clarify the mechanics of the calculations necessary to make probability statements. Given a mean annual rainfall of 664 mm, and a standard deviation of 120 mm, what is the probability of less than 500 mm occurring in any one year? Assuming a normal distribution, the problem is illustrated in Fig. 21A, where the shaded area represents the required probability value. The stages involved in the calculations are set out below:

(a) The rainfall value of 500 mm is $(664 - 500) = 164$ mm below the mean.
(b) 164 mm is equivalent to $\frac{164}{120} = 1.37$ standard deviations below the mean.
(c) The required z in Table A is therefore 1.37.

(d) The corresponding probability value in Table A is 0.9147 (or 91.47%).
(e) The required probability is $(1.0 - 0.9147) = 0.0853$ (or 8.53%). Note in particular that it is the small area in Fig. 21A that is required here.

FIG. ㉑ MAKING PROBABILISTIC STATEMENTS IN REALITY

A   THE PROBABILITY OF LESS THAN 500 MM OF RAINFALL
     AT EDINBURGH, SCOTLAND, IN ANY ONE YEAR

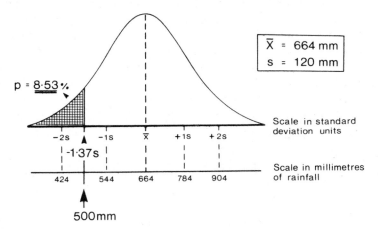

B   THE RAINFALL AMOUNT THAT IS EXCEEDED
     WITH A PROBABILITY OF 90% IN ANY ONE YEAR
     AT EDINBURGH, SCOTLAND.

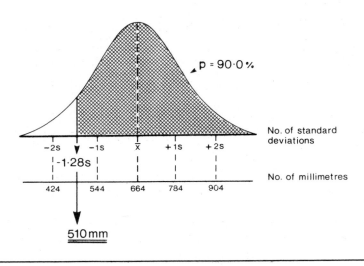

In the example above, we have answered a question of the form: 'What is the probability of rainfall exceeding (or being less than) a certain amount in any one year?' It may be more useful in some circumstances to answer the reverse kind of question, such as: 'What amount of rainfall has a probability of occurrence of greater than (or less than) a certain probability value, in any one year?' In the context of Edinburgh rainfall: what amount of rainfall can be relied upon to occur 9 years out of any 10? The stages are set out below and illustrated in Fig. 21 B:

(a) 9 years in 10 is equivalent to 90 years in 100 or a probability of 90%.
(b) The required p in Table A is therefore 0.90.
(c) The corresponding value of z is 1.28 in Table A.
(d) 1.28 standard deviations from the mean is equivalent to $(1.28 \times 120) = 154$ mm.
(e) 154 mm below the mean is a rainfall amount of $(664 - 154) = 510$ mm, which is the required amount of rainfall.

Probability maps

The usefulness of statistics to Geography is nowhere more obvious than in the probability map, in which the concept of probability is combined with the geographer's most important tool. The kind of calculations carried out earlier in this chapter, on data from particular places, can quite readily be performed on data relating to a large number of locations, and the results summarized in the form of a map.

In Fig. 21 A, the probability of receiving greater or less than a given amount of rainfall at one meterological station was calculated. Similar calculations applied to a large number of meterological stations would yield a set of probability values. In the same way that mean annual rainfall maps (Fig. 17) are based on mean annual rainfall values for individual stations, so probability values can be plotted and a generalized map produced. Figure 22 A depicts the probability of receiving greater than 750 mm of rainfall in any one year for the British Isles. Thus most of Highland Britain is likely to experience this amount of rainfall, or more, in 9 years out of 10 (probability 90%), whereas in eastern England, between the Humber and Thames estuaries, this amount of rainfall is unlikely to be reached in at least 7 years out of 10.

A second type of probability map is produced if amounts of rainfall are mapped, which correspond to a known probability of occurrence. Such a map involves the sort of calculation (for each meterological station) illustrated in Fig. 21 B. An example of this type of map is shown in Fig. 22 B, which indicates the amount of rainfall that is likely to be exceeded in 9 years out of 10, or with a probability of 90%. Thus over most of Highland Britain there is a 90% probability of at least 750 mm of rainfall in any one year (compare with the statement in the previous paragraph referring to Highland Britain).

Exercise 6. Use of tables of the normal distribution function for making elementary probability statements.

Background

This exercise has two parts. Question 1 concentrates on the manipulation of tables of the normal distribution function (Table A, Appendix). The remaining questions all require similar manipulations in the context of actual data. These questions are all extensions of previous exercises; they involve the concept of probability in problems that were considered previously in a deterministic manner.

Practical work (a normal curve should be drawn in connection with each answer, showing the area required, as in Fig. 21)

1. Using Table A, calculate the following:
   (a) The probability of any one measurement in a set of measurements being above one standard deviation below the mean (i.e. $> -1s$).
   (b) The probability of any one measurement being below two standard deviations below the mean (i.e. $< -2s$).
   (c) The probability of any one measurement lying between three standard deviations above and below the mean (i.e. between $\pm 3s$, or $< +3s$ but $> -3s$).
   (d) The probability of any one measurement being above one standard deviation below the mean. but below two standard deviations above the mean (i.e. $> -1s$ and $< +2s$).
   (e) How many standard deviations above the mean must a value lie in order to be 90 % ($p = 0.9$) certain that any one measurement will be less than (or will not exceed) that value?
   (f) How many standard deviations below the mean must a value lie in order to be 95 % sure that any one measurement will exceed that value?
   (g) What is the probability of any one measurement being outside of three standard deviations from any mean?

2. Using your answers to question 1, Exercise 5, relating to monthly rainfall at Cardiff, calculate the following:
   (a) The probability of greater than 40 mm of rain in any one June.
   (b) The probability of less than 50 mm of rain in any one December.
   (c) The probability of less than 150 mm in any one December.
   (d) The probability of between 40 mm and 100 mm in any one June.
   (e) The minimum amount of rain expected in June with a probability of 95 %.
   (f) The amount of rain that you would expect to be exceeded in June in 19 years out of any 20.
   (g) If a drought is experienced in any June with less than 20 mm of rain, what is the probability of a drought in any one June?
   (h) In any period of 100 years, how many droughts are likely in June?
   (i) If flooding is experienced in a December in which greater than 150 mm of rain falls, what is the probability of a flood next December?
   (j) State at least two statistical assumptions made in (a)–(i).
   (k) State at least two meteorological/hydrological limitations to the answers in (g)–(i).

3. In Fig. 16 A and B, the quality of life index for black census tracts in Atlanta, Georgia, U.S.A., was shown by the use of dispersion diagrams to be lower than the quality of life index for white census tracts. The mean and standard deviation for the black and white tracts are:

   Black census tracts, mean $= -64.22$, standard deviation(s) $= 76.42$

   White census tracts, mean $= +78.04$, standard deviation(s) $= 107.66$

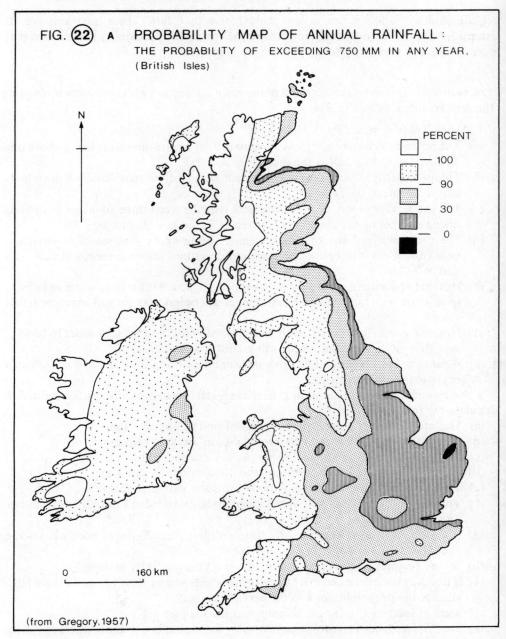

FIG. 22 A   PROBABILITY MAP OF ANNUAL RAINFALL :
THE PROBABILITY OF EXCEEDING 750 MM IN ANY YEAR,
(British Isles)

PERCENT
— 100
— 90
— 30
— 0

0   160 km

(from Gregory, 1957)

(a) What is the probability that a black census tract, selected at random, will possess a standard of living index that is higher than the mean value found in white tracts?

(b) What is the probability that a white census tract, selected at random, will possess a standard of living index that is lower than the mean value found in black tracts?

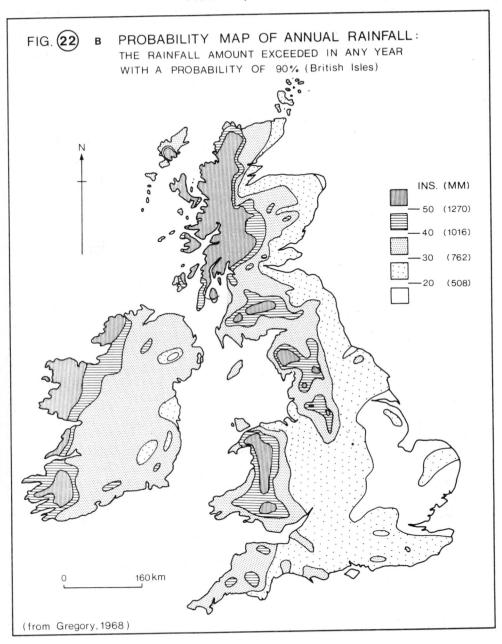

FIG. (22) B   PROBABILITY MAP OF ANNUAL RAINFALL:
THE RAINFALL AMOUNT EXCEEDED IN ANY YEAR
WITH A PROBABILITY OF 90% (British Isles)

INS. (MM)

— 50 (1270)

— 40 (1016)

— 30 (762)

— 20 (508)

0   160 km

(from Gregory, 1968)

(c) Calculate the minimum expected standard of living index for 95% of white tracts.
(d) Calculate the minimum expected standard of living index for 95% of black tracts.
(e) Discuss the limitations on using these data and answers as a test of the hypothesis that 'black households are grossly underprivileged in the U.S.A.'.

Exercise 7: Construction of a probability map of atmospheric pollution for north-west Europe.

Background

The chemical composition of the atmosphere can be modified by human activity, sometimes in extensive and complex ways. One method of monitoring the effect is to analyse the composition of rain-water and compare this with 'unpolluted' rain. A good indicator of degree of pollution is the acidity of rainfall, which is measured in pH units (a pH of 7.0 being neutral, and the lower the pH the higher the acidity).

Under natural, unpolluted conditions rain-water is normally weakly acid in reaction (pH about 6.0), due to the presence of carbon dioxide (CO_2) in the atmosphere. One of the major atmospheric pollutants is sulphur dioxide (SO_2), which forms a much stronger acid when in combination with atmospheric moisture. Under polluted conditions, therefore, rainfall becomes air-borne acid.

Figure 23 A and B shows the geographical distribution of acidity in precipitation over north-west Europe in 1956 and in 1966, an interval of 10 years. The values mapped are annual means of pH. While these maps indicate where acidity is greatest, on average, they give no information on the acidity that can be expected to occur at a known level of certainty, or how likely it is that rainfall will be more or less acid at any one time. Given a knowledge of the standard deviations associated with the mean values given in Fig. 23, it is possible to construct maps containing such probabilistic information.

Practical work

The data in Table 5 show the mean and standard deviation of the pH of precipitation at 48 sites in north-west Europe. The mean values were taken from the maps in Fig. 23; the standard deviations are invented values, not measured values.

1. The aim is to construct probability maps (one for 1956 and one for 1966) showing *the probability that a pH value of less than 6.0 will occur in any one fall of rain.* In effect, the maps will show the probability of any one fall being more acid than the natural background level of acidity.
   (a) For each of the sites listed, calculate and tabulate the probability of a pH value of less than 6.0 in any one fall.
   (b) Transfer the tabulated values to the map in Fig. 24.
   (c) Draw isolines of equal probability ('isoprobs'), interpolating sensibly, choosing a suitable interval for the isolines, and only entering isolines in areas where good control is available.

2. Using your probability maps (and Fig. 23) describe the major patterns and changes in rainfall acidity that have occurred between 1956 and 1966.

3. Offer some explanations for the pattern and changes including the following:
   (a) The areas of highest acidity.
   (b) The negligible change in acidity in south-west England.
   (c) The change in the pattern and intensity of acidity in the Scandinavian peninsula.
   (d) The change in Iceland.

4. Discuss the limitations of this application under the following headings:
   (a) Limitations of the data as an index of atmospheric pollution.

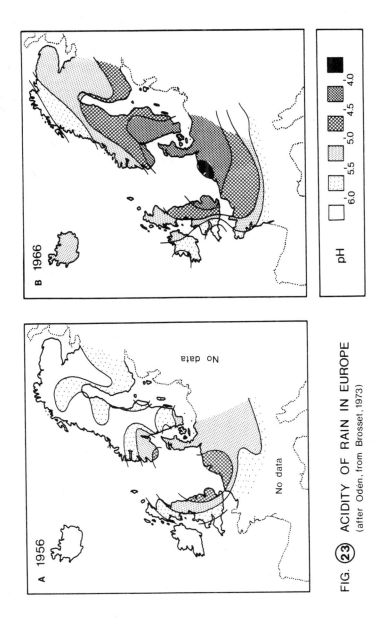

FIG. ㉓ ACIDITY OF RAIN IN EUROPE
(after Odén, from Brosset, 1973)

TABLE 5. *Acidity of rain in north-west Europe*

Site	Mean (1956)	Standard deviation (1956)	$\bar{x}$ (1966)	$\hat{\sigma}$ (1966)
1. Belfast	6.0	0.9	5.5	1.0
2. Galway	6.3	0.8	6.1	0.8
3. Dublin	6.1	0.7	5.7	0.9
4. Cork	6.2	0.9	5.9	0.9
5. Edinburgh	5.4	1.2	4.9	1.2
6. Manchester	5.4	1.1	4.7	1.1
7. Hull	4.9	1.4	4.6	1.3
8. Birmingham	5.3	1.4	4.8	1.2
9. Cardiff	5.7	1.0	5.6	1.1
10. Southampton	5.5	1.1	4.8	1.2
11. London	4.9	1.3	4.4	1.3
12. Brest	Not available		6.0	0.9
13. Nantes	Not available		5.3	1.2
14. Lyon	Not available		5.5	1.0
15. Paris	5.4	1.1	4.7	1.3
16. Dijon	Not available		4.8	1.3
17. Bern	5.4	1.1	4.9	1.2
18. Munich	5.4	1.1	4.9	1.3
19. Strasbourg	5.3	1.3	4.7	1.3
20. Frankfurt	5.1	1.3	4.4	1.3
21. Cologne	4.8	1.4	4.2	1.4
22. Brussels	4.8	1.5	3.9	1.6
23. Antwerp	4.7	1.5	3.8	1.5
24. Rotterdam	4.7	1.6	3.8	1.7
25. Amsterdam	4.7	1.5	3.9	1.6
26. Bremen	5.2	1.5	4.2	1.5
27. Hannover	5.2	1.5	4.2	1.5
28. Hamburg	5.2	1.6	4.3	1.5
29. Esbjerg	5.1	1.4	4.2	1.6
30. København	5.3	1.4	4.3	1.6
31. Malmo	5.4	1.5	4.4	1.5
32. Göteborg	5.4	1.3	4.4	1.5
33. Norrköping	5.8	1.2	4.5	1.4
34. Stockholm	5.8	1.1	4.6	1.3
35. Turku	5.7	1.0	4.7	1.1
36. Helsinki	5.7	1.0	4.8	1.2
37. Karlstad	5.8	1.3	4.4	1.4
38. Oslo	5.3	1.2	4.5	1.4
39. Kristiansand	4.9	1.2	4.4	1.3
40. Stavanger	5.0	1.1	4.7	1.2
41. Bergen	5.4	1.0	5.1	1.0
42. Trondheim	6.1	0.8	5.2	1.0
43. Sundsvall	6.1	0.9	4.7	0.9
44. Luleå	5.9	1.0	4.9	1.3
45. Narvik	6.2	0.9	5.8	1.1
46. Vadso	6.3	0.7	5.4	1.1
47. Hammerfest	6.2	0.9	6.0	1.0
48. Reykjavik	6.2	0.8	5.1	1.3

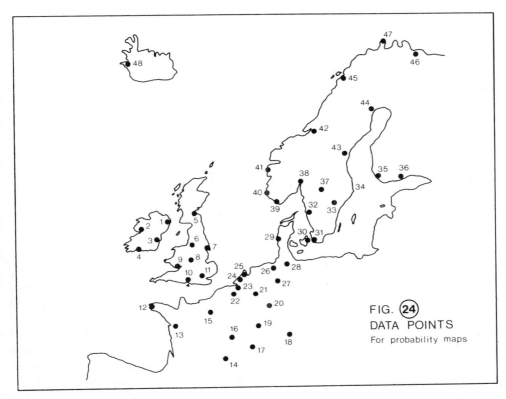

FIG. (24)
DATA POINTS
For probability maps

(b) Limitations of the calculation of probabilities at any site.

(c) Limitations of map construction from the values for all sites.

5. Describe and explain how you would go about constructing a different type of probability map from the same data, illustrating the kind of calculation necessary with reference to the mean and standard deviation for Cork in 1956 and in 1966.

5

The Problem of Time-dependence and Time-series Analysis

IN THE discussion of probability in the preceding chapter, it was assumed that the data were *representative samples* and that they were *normally distributed*. A third assumption was also made, namely that the individual measurements were *statistically independent*. In the context of annual rainfall, for example, it was assumed that a particular annual total was equally likely to occur in any of the years; that is, a particular annual amount was said to be independent of the amount received in the previous years and of the amount received in the years following. It is known, however, that climates often change gradually and that groups of wet or dry years sometimes occur. Under these circumstances climatic data are time-dependent. Such *time-dependence* is a common feature of data collected over any time interval; examples from the field of Human Geography include variations in agricultural production, industrial output and population growth.

Any *time-series* (a set of measurements collected over a period of time) may or may not exhibit time-dependence. Time-dependence can take a variety of forms: there may be *long-term trends* (when values tend to rise or fall gradually); there may be *irregular fluctuations* (when values tend to rise and fall over time and may return to approximately the same level), and there may be *cyclic fluctuations* (when similar fluctuations tend to repeat themselves through time). Figure 25A, a graph of the level of carbon dioxide in the atmosphere measured in Hawaii from 1958 to 1975, shows a time-series with pronounced trend and cyclic components. Unpredictable or *random variation* often obscures such patterns, however, as shown in Fig. 25B, which gives the annual rainfall totals in Edinburgh, Scotland, from A.D. 1785 to 1973 (one of the longest rainfall records in the world).

If the random component of a time-series is the only component, then it may be valid to apply the normal distribution function to the data in the manner described in the preceding chapter. If, on the other hand, trends, fluctuations or cycles are present, then the probability of a given measurement value cannot be calculated with reference to the normal distribution function, because the actual probability is statistically dependent on position in the series. This is a rather important limitation if one is using a set of data to predict likely values for future years.

One method of time-series analysis, which permits the detection of trends and non-random fluctuations, is the use of *running means* (also known as moving averages). A 5-

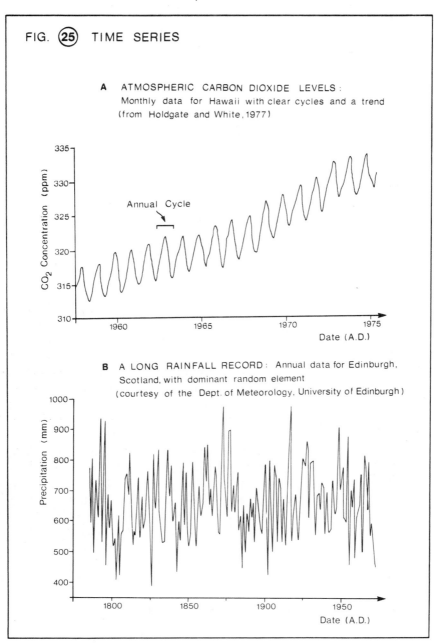

FIG. (25)   TIME SERIES

A   ATMOSPHERIC CARBON DIOXIDE LEVELS :
Monthly data for Hawaii with clear cycles and a trend
(from Holdgate and White, 1977)

Annual Cycle

B A LONG RAINFALL RECORD : Annual data for Edinburgh,
Scotland, with dominant random element
(courtesy of the Dept. of Meteorology, University of Edinburgh)

year running mean of river discharge data, recorded at a gauge on the Bristol Avon at Bath in south-west England, is shown in Fig. 26. The annual data plotted are the number of times that the river exceeded a discharge of 56.6 cumecs (cubic metres per second). The bold line is the 5-year running mean, and it shows a rising trend which reflects the increasing frequency of floods in recent years.

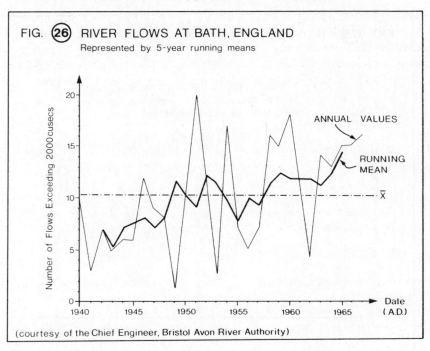

FIG. (26) RIVER FLOWS AT BATH, ENGLAND
Represented by 5-year running means

(courtesy of the Chief Engineer, Bristol Avon River Authority)

The principle of this approach lies in a property of the mean. If one calculates the mean of a number of measurements then the value obtained is a central tendency and averages-out or masks the high and low values. A running mean simply involves the calculation of overlapping means. The first mean plotted in Fig. 26 (to the left) is a mean of the number of discharges recorded in the first 5 years of the record (that is, 1940–4). The second value plotted is the mean for the period 1941–5, and so on, to the last mean for 1963–7. Each mean overlaps the next for 4 years, and each mean is plotted opposite the middle year of the 5 year period involved. The superimposed graphs in Fig. 26 illustrate the smoothing effect produced by use of a running mean, an effect resulting from the smoothing effect of each individual mean on the five individual measurements within it. Note also that the running mean graph is 4 years shorter than the graph of the annual measurements. This loss in length of the series is inevitable when running means are used and depends on the number of measurements involved in each calculated mean. The more measurements included in each mean, the shorter the length of the running mean graph, and the smoother are its fluctuations.

Another property of the running mean (which can be an advantage or a disadvantage, depending on the context) is that the number of measurements involved in each calculated mean determines the length of any non-random fluctuation that can be detected. Where a 5-year running mean is used, as in Fig. 26, any fluctuation that lasts for less than 5 years will be masked. In general, fluctuations lasting for a greater period than the number of years involved in the calculation of each mean will be registered in a running mean graph. The logical conclusion of extending the number of years included in each mean is to take the mean of the whole series, which results in the elimination of all fluctuations of whatever length (Fig. 26).

A detailed time-series analysis might use a number of running mean graphs in order to detect fluctuations of differing frequency and amplitude (length and height) by application of different degrees of smoothing. Another possible extension of the technique is to use *weighted means*, in which the middle years in each mean could be given greater weight than the outlying years. An example of a weighted 5-year mean is given by:

$$\frac{x_1 + 2x_2 + 3x_3 + 2x_4 + x_5}{9}$$

where x_1 to x_5 are five individual measurements. A running mean based on such weighted means would smooth the data in a different way to unweighted means.

Exercise 8. Running means and the analysis of birth and death rates in pre-industrial Norway.

Background

A general theory proposes that during the development of any country, population growth passes through a number of stages. Initially, population growth is slow, and characterized by high birth and high death rates. With the onset of 'development', particularly medical facilities, death rates (particularly infant mortality rates) are rapidly reduced. Finally, but after a considerable time-lag, birth rates are also reduced, a condition represented by advanced Western societies. This theory states, therefore, that population growth passes through a 'demographic transition' (during which rapid population growth occurs) between two stages during which birth and death rates are maintained in relative balance with only slow overall growth in the population. This theory or model is illustrated graphically in Fig. 27.

Very complete records of population growth are available for the Kingdom of Norway from the early eighteenth century. The registration of births and deaths was the responsibility of the State Church from A.D. 1687, and the first complete population

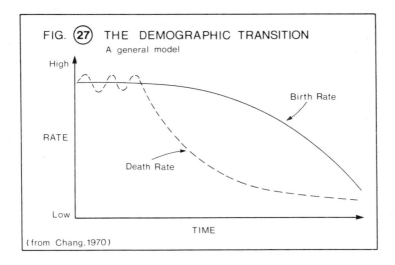

FIG. 27 THE DEMOGRAPHIC TRANSITION
A general model

Birth Rate

Death Rate

RATE

High

Low

TIME

(from Chang, 1970)

census took place in 1769. These and similar data from neighbouring countries of northern Europe are important because they provide rare examples of long series of measurements relating to population growth from pre-industrial times. In many ways, population fluctuations in Norway in the eighteenth and nineteenth centuries reflect the form of population growth in underdeveloped countries at present. The Norwegian data provide, therefore, a useful test of the general demographic transition model.

These data also enable the investigation of the relationship between population growth and the physical environment, a relationship which has changed as the country has 'developed'. Running means are used in this exercise to analyse data on birth rates and death rates in Norway from 1735 to 1855. In this way, major trends and fluctuations are detected, and the changing relationship with the physical environment can be appreciated.

Practical work

1. The data given in Table 6 are birth and death rates for Norway from A.D.1735 to 1855. The rates are expressed as the number of births per 1000 population and the number of deaths per 1000 population, respectively.

TABLE 6. *Birth and death rates in Norway (1735–1855)*

Year	Births	Deaths	Year	Births	Deaths	Year	Births	Deaths	Year	Births	Deaths
1735	29.0	19.0	1765	31.2	28.1	1795	32.3	22.5	1825	34.4	17.5
1736	30.3	20.5	1766	31.0	27.7	1796	31.7	21.7	1826	34.9	18.5
1737	30.1	24.5	1767	32.2	22.2	1797	32.8	22.5	1827	32.1	18.0
1738	27.6	22.8	1768	30.2	22.3	1798	32.3	22.6	1828	31.8	19.4
1739	30.4	22.8	1769	30.9	21.8	1799	32.6	21.0	1829	33.7	19.4
1740	28.8	25.1	1770	31.5	23.6	1800	30.0	25.6	1830	32.4	19.7
1741	26.8	40.8	1771	31.1	22.9	1801	28.3	27.4	1831	31.0	19.8
1742	25.7	52.2	1772	27.8	26.8	1802	27.2	25.2	1832	29.9	18.5
1743	27.7	28.4	1773	23.4	47.5	1803	29.1	24.9	1833	30.7	20.4
1744	29.5	21.4	1774	28.0	25.5	1804	27.4	23.5	1834	31.7	22.5
1745	31.9	18.2	1775	33.0	22.9	1805	30.1	20.7	1835	32.7	19.5
1746	29.1	20.6	1776	29.3	20.4	1806	30.5	21.1	1836	29.4	19.3
1747	32.0	23.2	1777	30.9	20.8	1807	29.7	22.6	1837	28.7	20.8
1748	31.8	32.1	1778	31.0	20.0	1808	27.8	26.1	1838	30.3	21.7
1749	32.0	27.1	1779	31.1	27.1	1809	22.3	35.9	1839	26.7	21.6
1750	29.7	25.5	1780	32.1	25.3	1810	26.9	26.8	1840	27.8	19.8
1751	33.9	26.1	1781	31.0	20.7	1811	27.7	25.5	1841	29.8	17.3
1752	32.4	24.6	1782	30.6	22.4	1812	29.5	21.3	1842	30.8	18.0
1753	33.7	22.7	1783	27.4	24.6	1813	26.1	29.5	1843	30.2	17.9
1754	34.2	23.4	1784	30.2	23.8	1814	24.5	22.6	1844	29.9	17.1
1755	32.6	24.8	1785	28.7	33.1	1815	30.6	19.8	1845	31.2	16.9
1756	35.0	26.3	1786	30.3	24.2	1816	35.2	19.4	1846	31.1	17.9
1757	33.5	21.5	1787	29.0	22.7	1817	32.5	17.7	1847	30.8	20.4
1758	32.6	23.9	1788	30.6	26.1	1818	30.8	19.1	1848	29.8	20.5
1759	31.5	26.2	1789	30.5	30.4	1819	32.0	19.7	1849	32.1	18.3
1760	34.2	22.6	1790	31.9	22.9	1820	33.4	18.9	1850	30.9	17.2
1761	33.0	22.6	1791	32.6	22.9	1821	34.8	20.5	1851	31.9	17.1
1762	32.9	23.2	1792	34.6	23.9	1822	33.0	19.5	1852	31.0	17.9
1763	31.4	35.6	1793	34.0	22.1	1823	34.0	17.8	1853	32.0	18.3
1764	32.9	27.0	1794	33.6	20.8	1824	32.5	18.5	1854	34.2	16.0
									1855	33.4	17.1

(From Drake, 1965.)

These time series are plotted as graphs in Fig. 28. The aim is to construct the corresponding graphs using *5-year running means*.
  (a) Calculate and tabulate the birth rate 5-year running means.
  (b) Calculate and tabulate the death rate 5-year running means.
  (c) Draw up superimposed graphs of the smoothed birth and death rates.
  (d) Note the smoothness of the running mean graphs compared to the graphs of the raw data.
  (e) Note the highest and lowest values and the amplitude (difference between the highest and lowest values) for all four graphs.

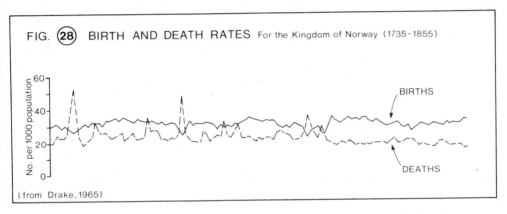

FIG. ㉘ BIRTH AND DEATH RATES For the Kingdom of Norway (1735–1855)

(from Drake, 1965)

2. Describe the graphs of birth and death rates, paying particular attention to the following:
  (a) Trends.
  (b) Non-random fluctuations.
  (c) Random variation.
3. Describe the main features of population growth, defining in particular periods when the total population of Norway is likely to have declined and periods during which the population of the country grew most rapidly.
4. What other factor(s) need to be taken into account (besides birth and death rate data) to determine the actual population change over a number of years?
5. Compare your running mean graphs with the 'demographic transition model' given in Fig. 27.
  (a) At what point in time (if any) can Norway be said to have entered the demographic transition? Fully justify your answer.
  (b) Do the Norwegian data suggest any modification to the model depicted in Fig. 27?
6. In the eighteenth century the Norwegian economy, especially in the north and east of the country, was dominated by grain cultivation. Grain harvest failures occurred over wide areas of the country in 1741–2, 1748, 1762–3, 1773, 1784–5, 1807–9 and 1812. The early years of the 1740s were particularly severe and were known as the 'green years', when grain failed to ripen and bread was made from lichens, birch bark and other plants. In one area straw was taken from dung heaps, washed, mixed with meal and baked. Reduced tree-growth and/or glacier advances indicate abnormally cold summers in the early 1740s, the early 1780s, the early 1800s, the early 1820s, about 1840, and about 1850.

Potatoes were introduced into Norway in the 1750s, and increased in cultivated area very sharply between 1800 and 1830 becoming the major crop in the country by 1835. During the Napoleonic Wars, importation of grain into Norway was prevented on several occasions between 1807–14 by the British fleet, at a time when about 25% of Norwegian needs were normally imported. In the first decade of the nineteenth century, herring and cod fisheries failed in coastal areas of the country. In the first 30 years of the nineteenth century, the harvest per head of the agricultural population increased by 70%. Between 1820 and 1865 the cultivated area of Norway doubled. Vaccination against smallpox was made compulsory by a royal edict in 1810 and was quickly adopted for children:

Years	Vaccinated as a percentage of	
	population	live-births
1802–10	0.43	15.6
1811–20	0.92	30.7
1821–30	1.59	47.9
1831–40	1.72	58.1
1841–50	2.31	75.3
1851–60	2.69	81.5

(After Malm, from Drake, 1965)

Using the above facts relating to possible causes of changes in birth and death rates, construct a reasonable explanation for the running mean graphs and population changes in Norway from 1735 to 1855. Your explanation must account for all the evidence and be internally consistent.

6

The Problem of
Non-normality and
Data Transformations

IT HAS been shown in earlier chapters that descriptive statistics, particularly the mean and standard deviation, are most successfully applied to sets of measurements that have a symmetrical distribution. It should also be clear that, assuming the data are statistically independent, it is possible to make inferences about populations based on a representative sample, if the measurements have a particular type of symmetrical distribution—namely a normal distribution. However, as pointed out in Chapter 2, geographical data are often statistically 'dirty', that is they do not conform to the ideal statistical model. One of the most important problems in statistical analysis is that data are often skewed. *Skewed distributions* are asymmetrical and are therefore more difficult to analyse in a meaningful way.

What are the possible solutions to non-normality? Two solutions are available. The first solution is to employ 'distribution-free' or *non-parametric statistics*, an approach that is being increasingly adopted by geographers. The validity of non-parametric statistics does not depend on symmetry or a normal distribution. Nevertheless, they do have limitations, and in particular require interval scale measurements to be 'degraded' to a lower-order scale of measurement (such as an ordinal scale). Some non-parametric inferential statistics will be treated later in the manual. The present chapter deals with the second solution to non-normality.

If a distribution is skewed it may be possible to *transform* the distribution to produce a normal shape. The solution is possible only if the distribution has one mode (uni-modal) as is illustrated in Fig. 8. An example of a positively skewed distribution is provided by the number of people in different income brackets. Many more people have low incomes than have high incomes. The distribution of incomes therefore tends to resemble Fig. 8A with relatively few people in the tail to the right-hand side of the distribution. Another example is provided by the age of plants on a patch of bare ground that has been subject to invasion and colonization for several years; there would be many young plants, but relatively few old ones. A forestry plantation, on the other hand, will tend towards a negatively skewed age-distribution. There is likely to be many relatively mature trees, no trees will be older than the original seedlings at the time of planting, and there will tend to be relatively few younger generation trees (survivors of forest management). The younger generations are represented by the tail to the left-hand side of the distribution in Fig. 8B.

How does one go about transforming a distribution like the positively skewed distribution in Fig. 8A? Consider the effect of 'stretching' the measurement scale in this figure. If the scale is 'pulled-out' such that the right-hand side of the distribution is stretched more than the left-hand side, the resulting distribution is even more skewed than before. If, on the other hand, the left-hand side is stretched more than the right-hand side, then it is possible to produce a normal distribution. Too little 'stretching', and normality will not be reached; too much 'stretching' and a positive skew will be transformed into a negative skew.

A commonly used method for the transformation of a positively skewed distribution involves taking the logarithms of each measurement in the data set. The effect is to bring high values relatively 'close' together (recall the arrangement of lines on log-paper). Regrouping of the measurements into class-intervals on the transformed scale produces a modified histogram, as shown in Fig. 29. This example depicts the number of plants (vertical axis) in different age-classes (horizontal axis) on ground cleared of vegetation 10 years previously. The actual age of each plant and the logarithms of these ages are as follows:

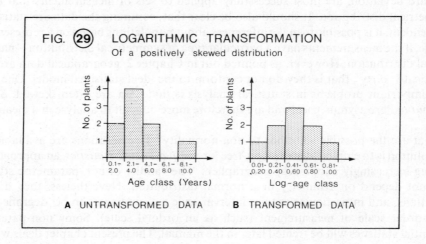

FIG. ㉙   LOGARITHMIC TRANSFORMATION
Of a positively skewed distribution

A UNTRANSFORMED DATA   B TRANSFORMED DATA

Age (years)	1.5	2.0	2.4	3.0	3.4	3.8	4.4	5.6	8.1
Log age	0.18	0.31	0.38	0.48	0.53	0.58	0.64	0.75	0.91

The untransformed data results in a positively skewed histogram when a 2-year age-class is used (Fig. 29A). The transformed data, with a class-interval of 0.2 log age units, shows that a successful transformation to normality has been achieved (Fig. 29B).

A logarithmic transformation is quite powerful. If a distribution is less skewed than in the example above, then a milder transformation (less stretching), such as a square-root transformation, can be used. Similarly, the transformation of a negatively skewed distribution could involve squaring the original values. Such a transformation 'stretches' the distribution in the opposite direction to a logarithmic transformation. Selection of an appropriate transformation requires, therefore, a consideration of the type and degree of skewness that is characteristic of a data set. A number of transformations are summarized below:

Type of skewness	Degree of skewness	
	Weak	Strong
Positive	$\sqrt{x}$	log x
Negative	x^2	antilog x

Once a normal distribution has been obtained, such as in Fig. 29B, it is then possible to begin answering questions of the kind: 'What is the probability of any one individual measurement being greater than a specified amount from the mean value?' To answer this question, the mean and standard deviation of the *transformed values* must be calculated. Then the number of standard deviations from this mean can be used in conjunction with tables of the normal distribution function.

The question remains, however, of how normal does a distribution have to be to enable valid use of the normal distribution function? At the present time geographers are usually prepared to accept rather wide departures from normality in their analyses, but they are rarely very precise about this point.

Use of probability paper

One way of assessing whether a data set is normally distributed is to draw a histogram or a frequency distribution curve. An even easier method of testing for normality is by use of probability paper. This is a special type of graph paper on which a normal distribution (plotted in a particular way) results in a straight line, whereas skewed distributions plot as curves.

Figure 30A shows a frequency histogram and a normal *frequency distribution curve*. Figure 30B shows the same data plotted as a *percentage frequency* distribution curve, the number of occurrences (vertical axis) having been converted to a percentage of the occurrences. Note that in this example there are ten individual measurements involved. *Cumulative percentage frequency* has been plotted as the vertical axis in Fig. 30C; the corresponding curve is sigmoidal (or 'S'-shaped) in form. In this figure, the percentage of the occurrences in a particular class *and below* has been plotted. Note also that the vertical axis is divided equally and regularly. If the same cumulative percentage frequencies are plotted on probability paper (Fig. 30D) a straight line results because, in this example, the original data (Fig. 30A) were normally distributed. Note that the only difference between the axes in Figs. 30C and D lies in the spacing or intervals on the vertical axis. Probability paper has this very useful property of representing normal distributions as straight lines.

Figure 31A shows the expected result when a positively skewed distribution is plotted in this way on probability paper. The cumulative percentage frequency curve is convex upwards. After a square-root transformation (suitable for a mild positive skew) a plot of the transformed data may give a straight-line graph (Fig. 31B). In this case, a logarithmic transformation (more powerful, and therefore more suited to a severe skew) would result in a curve that is concave upwards (Fig. 31C), indicating that the distribution has been converted to a negative skew; that is the data has been over-transformed (assuming the aim was to achieve a normal distribution).

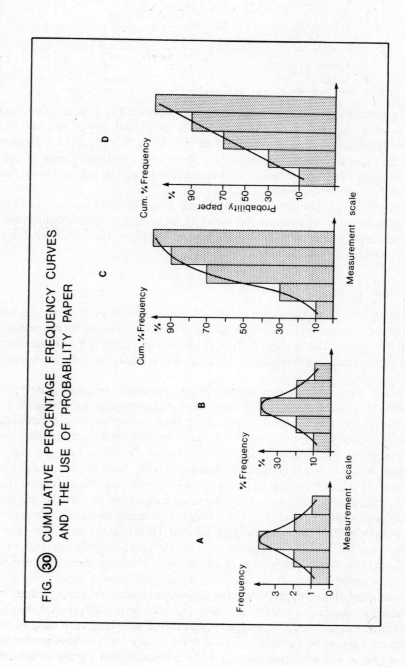

FIG. ㉚ CUMULATIVE PERCENTAGE FREQUENCY CURVES
AND THE USE OF PROBABILITY PAPER

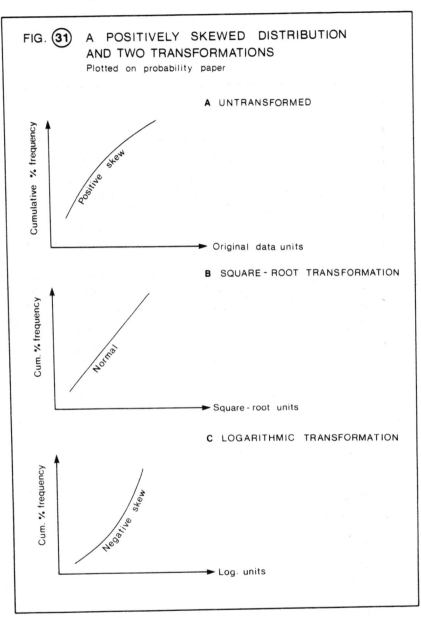

FIG. ③ A POSITIVELY SKEWED DISTRIBUTION AND TWO TRANSFORMATIONS
Plotted on probability paper

A UNTRANSFORMED

Cumulative % frequency

Positive skew

Original data units

B SQUARE-ROOT TRANSFORMATION

Cum. % frequency

Normal

Square-root units

C LOGARITHMIC TRANSFORMATION

Cum. % frequency

Negative skew

Log. units

The use of probability paper does not require the transformation and regrouping of individual measurements. The class mid-points can be taken from a histogram of the original data (stage 1 in Fig. 30A); transformed class mid-points can then be used with a transformed measurement scale (stages 2 to 4 in Fig. 30). In this way a number of transformations may be carried out rapidly, thus assuring that the best transformation for the data is identified. Once the best transformation has been found, then the individual measurements can be transformed in order to take the analysis further. The application of probability paper thus permits the visual identification of non-normality and the visual selection of an appropriate transformation. Its main advantage lies in the ease with which the straightness of a line can be judged visually; its main disadvantage lies in the possible influence of chosen class-intervals on results (a limitation of histograms also).

Exercise 9. Use of transformations and probability paper in the analysis of the wealth of nations at a world scale.

Background

The poverty of Africa and South-east Asia in particular is in marked contrast to the developed nations of North America and Europe. This gap between 'rich' and 'poor' nations is not closing despite the increasing amount of international aid to the 'Third World'. This exercise attempts to describe and summarize quantitatively some of the main disparities on a world scale, and should be concluded with a discussion of the causes of these disparities.

The data given in Table 7, and summarized in Fig. 32, are the *per capita* national incomes of all countries for which estimates have been made by the United Nations Statistical Office. No importance should be attached to small differences between nations, because some estimates are more reliable than others. Some estimates are based on reliable national sources, some are adjusted official figures, while others are rough estimates where no national figures have been published. All estimates have been converted into U.S. dollars using currency exchange values.

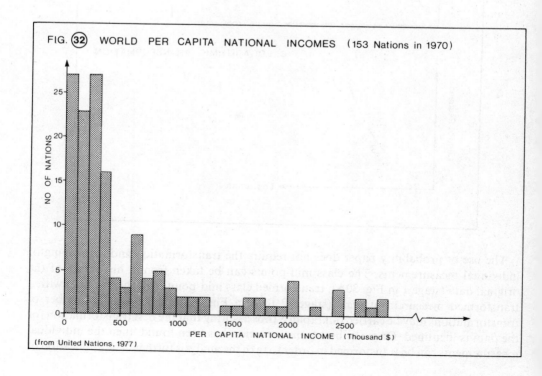

FIG. 32   WORLD PER CAPITA NATIONAL INCOMES (153 Nations in 1970)

NO OF NATIONS

PER CAPITA NATIONAL INCOME (Thousand $)

(from United Nations, 1977)

TABLE 7. *Per capita national incomes in 1970*

Country	$	Country	$	Country	$
AFRICA		**LATIN AMERICA**		**ASIA—E. & S.E.**	
Algeria	295	Antigua	332	Afghanistan	83
Angola	284	Argentina	984	Bangladesh	59
Benin	76	Barbados	618	Bhutan	44
Botswana	132	Belize	405	Brunei	1178
Burundi	60	Bolivia	175	Burma	73
Central African		Brazil	468	Dem. Kampuchea	119
Republic	119	Br. Virgin Isles	1190	East Timor	103
Chad	70	Chili	659	Hong Kong	747
Comoros	97	Colombia	310	India	94
Congo	213	Costa Rica	525	Indonesia	70
Egypt	202	Dominica	271	Japan	1636
Equat. Guinea	253	Dominican Rep.	334	Korea, Rep. of	250
Ethiopia	68	Ecuador	250	Lao People's Dem.	
Fr. Terr. of		El Salvador	281	Republic	71
Afars & Issas	1046	Grenada	332	Malaysia	345
Gabon	468	Guadeloupe	698	Maldives	86
Gambia	101	Guatemala	320	Nepal	73
Ghana	236	Guyana	323	Pakistan	163
Guinea	79	Haiti	94	Philippines	164
Guinea-Bissau	247	Honduras	266	Singapore	870
Ivory Coast	324	Jamaica	641	Sri Lanka	166
Kenya	127	Martinique	818	Thailand	167
Lesotho	91	Mexico	632		
Liberia	189	Montserrat	517	**EUROPE**	
Libyan Arab Rep.	1412	Neth. Antilles	1275	Austria	1730
Madagascar	127	Nicaragua	393	Belgium	2417
Malawi	66	Panama	646	Denmark	2898
Mali	53	Paraguay	239	Finland	1998
Mauritania	136	Peru	302	France	2490
Mauritius	223	Puerto Rico	1738	Germany, Fed. Rep.	2749
Morocco	221	St. Kitts–Nevis–		Greece	1090
Mozambique	216	Anguilla	248	Iceland	2058
Niger	81	St. Lucia	328	Ireland	1254
Nigeria	130	St. Vincent	222	Italy	1585
Reunion	769	Surinam	650	Luxembourg	2638
Rwanda	54	Trinidad and		Netherlands	2232
Senegal	219	Tobago	732	Norway	2458
Sierra Leone	150	Turks & Caicos		Malta	721
Somalia	87	Islands	380	Portugal	677
South Africa	662	Uruguay	809	Spain	985
S. Rhodesia	258	Venezuela	932	Sweden	3719
Sudan	109			Switzerland	3072
Swaziland	272	**ASIA—MIDDLE EAST**		United Kingdom	2031
Togo	125	Bahrain	888		
Tunisia	260	Cyprus	873	**OCEANIA**	
Uganda	127	Iran	352	Australia	2660
United Rep. of		Iraq	309	Fiji	385
Cameroon	179	Israel	1655	Fr. Polynesia	2001
United Rep. of		Jordan	261	New Caledonia	3079
Tanzania	94	Kuwait	2814	New Zealand	2030
Upper Volta	62	Lebanon	589	Papua New Guinea	255
Zaire	76	Oman	292	Solomon Islands	178
Zambia	365	Qatar	1837	Tonga	162
		Saudi Arabia	495	Western Samoa	177
NORTH AMERICA		Syrian Arab Rep.	259		
Canada	3366	Turkey	350		
United States	4285	Yemen	77		
		Yemen, Democrat.	92		

(From United Nations, 1977.)

The graphical representation of this data in the form of a histogram (Fig. 32) reveals a strong positive skew. Mean income values calculated from the data would be misleading descriptive statistics (either for the whole data set or for sub-sets of the data). Furthermore, probabilistic statements based on the mean and standard deviation would not be valid. The transformation of this distribution to normality is therefore an essential preliminary to this type of analysis.

Practical work

1. Using probability paper and the data arranged in classes (Fig. 32) plot the raw data in the form of a cumulative % frequency curve. Cumulative % frequency should be derived and tabulated in the following way:

Class interval (Income class)	Frequency (No. of countries)	% frequency	Cumulative % frequency
0–99	27		
100–199	23		
200–299	27		
300–399	16		
400–499	4		
etc.			

Cumulative % frequency is plotted against the class mid-points on probability paper.

2. Select two transformations appropriate for a positively skewed distribution, transform the *class mid-points*, calculate cumulative % frequency values for these transformed data, and plot curves on probability paper. Cumulative % frequency is plotted against the transformed class mid-points.

3. Describe and explain the three cumulative % frequency curves that you have plotted, paying particular attention to their degree of curvature and to the relative appropriateness of the transformations.

4. Describe and explain the form of cumulative % frequency curve that you would expect to result from the application of a transformation involving taking the square of the class mid-points.

5. Using the most appropriate transformation (determined in question 3), transform the individual *national values* of *per capita* income.
   (a) Calculate the mean and standard deviation of the untransformed data.
   (b) Calculate the mean and standard deviation of the transformed data.
   (c) Describe the differences between your answers in (a) and (b) and account for the differences.

6. Using the transformed mean and standard deviation derived in question 5(b), calculate:
   (a) The *per capita* national income that is likely to be exceeded by 90 % of nations.
   (b) The minimum expected *per capita* national income in the 'richest 10 %' of nations and the maximum expected in the 'poorest 10 %' of nations.
   (c) The probability that any one nation chosen at random will have a *per capita* national income lower than that in the United Kingdom.

(d) The proportion of nations that are expected to have a *per capita* national income that is higher than the United Kingdom value.

7. Using the transformed data for individual nations derived in question 5, calculate the mean and standard deviation for each of the following sub-sets of the data:

(a) Africa.

(b) Latin America.

(c) Asia.

(d) Europe.

8. Calculate the minimum expected *per capita* national income in the richest 10 % of each of these sub-sets and the corresponding values for the maximum expected in the poorest 10 %.

9. Using the results of your calculations in questions 6 to 8, describe the major differences in income at a world scale.

10. Write an essay on the possible causes of the disparities revealed in your analyses.

7

Elements of Sampling Methodology

SAMPLING methodology is concerned with the study of the methods that are available for obtaining a representative sample of a population. More precisely, it is concerned with the theory and practice of obtaining, as efficiently as possible, accurate *sample estimates* of *population parameters*. In the preceding chapters, several properties of samples have been discussed and it was assumed that they were accurate estimates of the corresponding characteristics of populations. The most important of these may be tabulated:

Sample mean	$\bar{x}$	Population mean	μ
Sample standard deviation	s	Population standard deviation	σ
Sample size	n	Population size	N

Samples are not always representative. Descriptive and inferential statistics will be of little use if they are summaries of non-representative samples. More generally, results can only be as good as the data on which they are based, and sampling methods are a major determinant of data quality.

What is involved in choosing a suitable sampling method? Five major factors are considered in turn below, each factor being viewed in relation to *efficiency* and *accuracy*. Efficient methods (requiring little cost/time/effort) are desirable but sufficient accuracy of the resulting estimate must be maintained.

(i) The *purpose* for which the estimates are required is obviously of major importance. Some purposes demand greater accuracy than others. For example, if a new drug or vaccine was to be tested by a sample survey of its effects on volunteers, then a very high level of accuracy would be necessary. If, on the other hand, a new detergent was to be put on the market and research was undertaken to assess the probable demand for the new product, a relatively inaccurate estimate would not be quite so disastrous; the greater the investment in the new product, however, the more important it would be to obtain a highly accurate prediction. Some methods are relatively inaccurate and designed for high efficiency but may be ideal for some purposes. Contrast, for example, a slope angle measured by an Abney level and a theodolite.

(ii) A second factor for consideration is the *sampling frame*. The sampling frame is the setting in which the *sampled population* is found. Ideally, sampling should be from the *target population* (about which information is required) but some sampling frames result in the sampled population being only a sub-set of the target population. For example, if one was estimating the characteristics of people in contrasting areas of a city by interview techniques, one possible sampling frame would be the telephone directory. The problem here is that not all people have a telephone and that a sample taken from the directory will under-represent certain social and economic groups. In other words, the sampled population (the telephone-owners) is not the target population (all persons in the city). Of particular relevance to geographers are *spatial sampling frames*, such as maps and aerial photographs. Spatial sampling frames differ from non-spatial ones in that location on the earth's surface influences the choice of individuals from the sampled population. If the purpose of a study involves aspects of location, such as the recognition of a distribution pattern, the calculation of areal coverage or the analysis of spatial variations, then a spatial sampling frame will be needed.

(iii) *Sampled individuals* and *attributes* must be defined. During sampling, observation and measurement is directed towards specific individuals within the population of similar individuals, and towards specific attributes (properties) of the individuals. For example, 'velocity' is an attribute of 'rivers' and 'glaciers' (individuals); 'depth', 'colour' and 'texture' are attributes of a 'soil profile' (individual); 'coverage', 'density' and 'productivity' are attributes of 'plant species' (individuals). *Operational definitions*, which define exactly what is to be measured and exactly how sampling is to proceed, form therefore an essential part of any sampling scheme. Such definitions must enable the recognition, description and recording of individuals and attributes without ambiguity. In these ways measurement and sampling will be repeatable and unbiased.

(iv) Rules for selecting individuals from the sampling frame constitute the *sampling design*. Purposive or *judgement sampling* involves selection based on the opinion of whosoever is doing the sampling (the operator). This approach can be very efficient, particularly if an expert is involved, and is especially useful in preliminary and extensive studies. 'Typical', 'best' and 'representative' samples (particularly case studies) are often merely judgement samples. The major limitation of this kind of sampling is that there is no way of knowing the extent to which bias, such as personal prejudices, or circular argument, is entering into the scheme and into the results. Nevertheless, much useful information will continue to be accumulated based on judgement sampling and, in almost every case, choice of 'the study area' in geographical research projects will remain purposive.

The alternative approach to judgement sampling is *probability sampling*, which is defined as a procedure by which individuals are selected from the sampling frame such that each individual has a known chance of being in the sample. Three simple but important types of probability sampling designs are: *random, systematic* and *stratified* designs. These may be spatial (locational) or non-spatial (non-locational), depending on the sampling frame. Figure 33 shows examples of the spatial variety; the points in the figure could represent points for the sampling of vegetation on an aerial photograph or for the sampling of farms from a map. Lines (transects or traverses) or areas (quadrats or plots) are valid alternatives to points for particular purposes or for particular types of individual.

The random design in Fig. 33A was derived in the following manner. The two edges of the area within which the sample was to be taken were scaled-off and used as axes. Each individual point was then located by two random numbers, each being a coordinate of the point (note the broken lines in the figure). Random numbers are conveniently prepared in the form of *random number tables* from which numbers can be read off in any direction and within which every number has an equal chance of occurrence (see Table B, Appendix). It must be emphasized that a truly random sample has a precise statistical meaning and should not be confused with samples that are chosen in a haphazard manner. For example, some biogeographers falsely believe that a random sample of vegetation can be obtained by standing in the centre of the area of interest and throwing quadrats over their shoulder! The advantage of a truly random sample is that every individual in the sampling frame has an equal chance of selection, with the result that no known bias is influencing the sample. A comparatively minor disadvantage of a random sample is that for some geographical purposes an even distribution of points is an advantage, and a random sample is certainly not evenly distributed over space.

The systematic design (Fig. 33B) is characterized by a regular sampling interval. In non-spatial context this might involve sampling of every tenth individual; in the two-dimensional spatial case regularity might be achieved by sampling at the intersections of a grid, as in Fig. 33B. This design has the advantage of an even coverage, which means that spatial variation within the area of interest is likely to be efficiently sampled. Its main disadvantage is that regularity within the phenomenon being sampled may clash with the regularity of the design and hence result in a biased sample. For example, if an area of bog hummock/hollow or an area of stone stripes were sampled systematically, sampling might result in the selection of only one of the phases in the pattern. The systematic sampling design employed at most weather stations (systematic sampling through time at 9.00 am each day) can lead to a biased sample of some attributes of the weather, namely those attributes that possess a diurnal rhythm.

Stratified designs (Fig. 33C) utilize prior knowledge to subdivide the sampling frame, usually with a view to controlling the effect of certain independent factors. A sample is then taken from each subdivision. Such a procedure can be much more efficient than sampling from the sampling frame as a whole. The validity of this type of design depends, of course, on the reliability of the prior knowledge and its relevance to the purpose of the investigation.

There are many more varieties of sampling design but most, on close inspection, are combinations of the above-mentioned types and therefore possess a mixture of their advantages and limitations. It would be profitable to examine some of the work described in recent geographical periodicals, paying particular attention to sampling designs.

(v) Lastly, *sample size* (or the number of individuals selected from the sampling frame) must be considered. The bigger the sample size, the more accurate will be the sample estimate of a population parameter. The problem is to be efficient but to take sufficient. Figure 34 shows a graph that illustrates the gradual approximation of a sample estimate to a population parameter as sample size increases. In the figure, the sample estimate could be a sample mean approaching the population mean (the latter indicated by the horizontal broken line). The accuracy required in a particular study will be dependent on its purpose. Thus, in theory, there is a minimum sample size necessary to achieve sufficient accuracy (that is, necessary to ensure that the sample estimate will be sufficiently close to the line). A sample size that is smaller than this minimum cannot be relied upon to be representative. Small sample sizes are one of the most important reasons for the inaccurate predictions of election results from opinion polls.

One other aspect of sample size will be considered. *Non-sampling errors*, such as measurement precision and operator error (errors resulting from different workers interpreting and carrying out instructions differently), may prevent high accuracy being

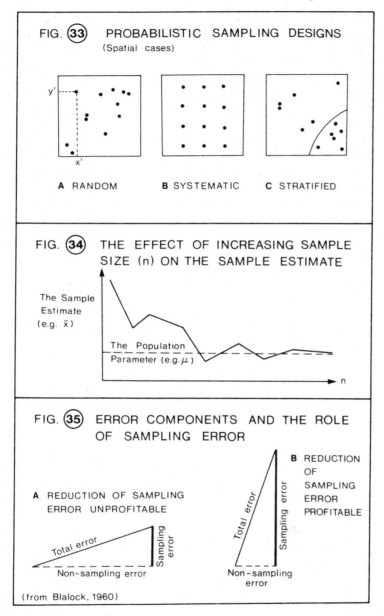

FIG. (33)   PROBABILISTIC SAMPLING DESIGNS
(Spatial cases)

A RANDOM   B SYSTEMATIC   C STRATIFIED

FIG. (34)   THE EFFECT OF INCREASING SAMPLE
SIZE (n) ON THE SAMPLE ESTIMATE

The Sample
Estimate
(e.g. $\bar{x}$)

The Population
Parameter (e.g. μ)

n

FIG. (35)   ERROR COMPONENTS AND THE ROLE
OF SAMPLING ERROR

B REDUCTION
OF
SAMPLING
ERROR
PROFITABLE

A REDUCTION OF SAMPLING
ERROR UNPROFITABLE

Total error

Sampling error

Total error

Sampling error

Non-sampling error

Non-sampling
error

(from Blalock, 1960)

obtained, no matter how large the sample size. Figure 35 represents total error as the resultant of sampling error and non-sampling error. In Fig. 35B, an increase in sample size, with consequent reduction in sampling error, will have a relatively great effect on the total error, because non-sampling error is relatively small. Figure 35A, however, shows a situation where non-sampling error is relatively large; in this situation, a reduced sampling error produced by increasing the sample size will have little effect on the total error. A profitable increase in overall accuracy will, therefore, only be achieved by increasing the sample size in the case of Fig. 35B.

Exercise 10: Application of sampling techniques in the extraction of information from soil maps in north Wales.

Background

Figure 36 is a map of the soils around Rhyl, a coastal area of north Wales. The majority of the soils in this area have been classified into six types, based on the field work and classification of the Soil Survey of England and Wales. This exercise illustrates some of the problems that will be encountered in any attempt to obtain and use reliable samples from this map. The first problem may be phrased in the form of a question: 'What sample size is necessary to obtain a reliable estimate of the areal coverage of each soil type on the map?' The second problem involves the choice of a sampling design: 'What are the differences in practice between the results obtained from random sampling and the results obtained from systematic sampling?'

The remainder of the exercise considers the use of sampling to compare maps and to approach the testing of hypotheses. Figure 37 is a map of the underlying parent materials in the same area as Fig. 36, based on the Drift Geology recorded by the Geological Survey. A subjective appraisal of the two maps suggests that there is some correspondence between the two sets of patterns. This in turn suggests that the underlying parent material is an important determinant of soil type in the area mapped. There are, however, other soil-forming factors—such as relief, organisms (vegetation and animals including man), climate and time—which are independent of parent material and may account for some of the pattern of soil types in Fig. 36. It would be of some value, therefore, to obtain objective information on the extent to which particular soil types are associated with particular parent materials. In this way, application of sampling techniques may be considered the first stage in the testing of a hypothesis by objective procedures.

Practical work

1. The aim of this section is to draw a random sample of points from Fig. 36, using a number of sample sizes.
    (a) Construct a grid-reference system along the horizontal and vertical axes of Fig. 36. The interval must be suitable for use with random number tables.
    (b) Using the random number tables in Table B (Appendix), locate 80 random points on the map. After the 5th, 10th, 15th, 20th, 25th, 30th, 40th, 50th, 60th, 70th and 80th points have been located, calculate the percentage areal coverage of the following:
        (i) gleyed soils,
        (ii) brown earths,
        (iii) calcareous soils,
        (iv) podzolized soils.
    (c) Draw up graphs of estimated percentage cover (vertical axis) against sample size (horizontal axis) for these four soil types.

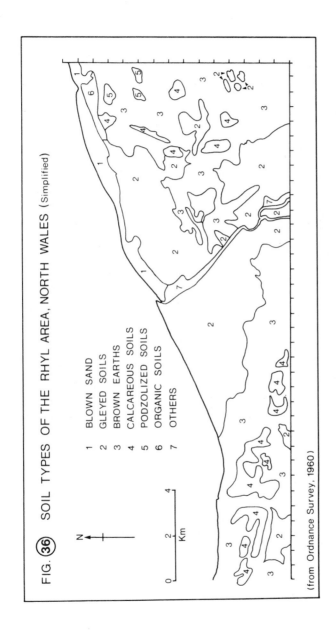

FIG. (36) SOIL TYPES OF THE RHYL AREA, NORTH WALES (Simplified)

1 BLOWN SAND
2 GLEYED SOILS
3 BROWN EARTHS
4 CALCAREOUS SOILS
5 PODZOLIZED SOILS
6 ORGANIC SOILS
7 OTHERS

(from Ordnance Survey, 1960)

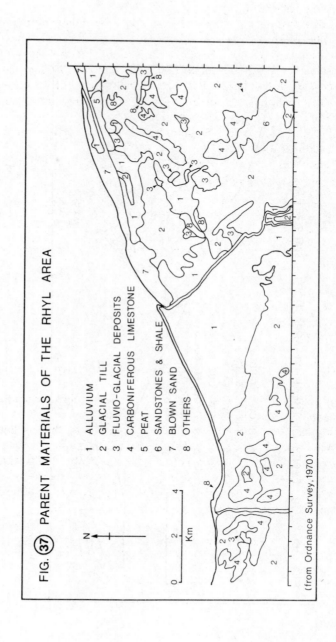

FIG. (37) PARENT MATERIALS OF THE RHYL AREA

1 ALLUVIUM
2 GLACIAL TILL
3 FLUVIO-GLACIAL DEPOSITS
4 CARBONIFEROUS LIMESTONE
5 PEAT
6 SANDSTONES & SHALE
7 BLOWN SAND
8 OTHERS

(from Ordnance Survey, 1970)

(d) Comment on the form of the graphs and draw conclusions about the necessary sample size for a representative sample in each case.

2. The aim of this section is to compare samples drawn by means of random and systematic sampling designs.

(a) By means of a grid, locate a systematic sample of 80 points on the map (Fig. 36).

(b) Compare the estimated percentage cover of the four soil types obtained by systematic sampling with the results obtained in (1) with a similar sample size.

(c) Draw conclusions about the relative suitability of the two alternative sampling designs, bearing in mind the differences to be expected on theoretical grounds.

3. The aim of this section is to compare the map of soil types with the map of parent materials (Fig. 37).

(a) Using either the random or the systematic design, transfer the same points to the parent material map.

(b) Cross-tabulate soil type with parent material. That is, enter the number of points falling in each cell of the table:

Soil Type	Parent Material Category				
	Alluvium	Till	Sand & Gravel	Limestone	Peat, etc
Blown sand					
Gleyed					
Brown earths					
Calcareous					
Podzolized					
Organic					
Other					

(c) Construct a similar cross-tabulation showing the *percentage* of each soil type within each parent material category.

(d) Construct a third cross-tabulation showing the percentage of each parent material category that is associated with particular soil types.

(e) Do the results in the three tables support the hypothesis that particular soil types are characteristic of particular parent materials? Fully justify your answer.

(f) Using your knowledge of the properties of the soils and of the parent materials, suggest some ways in which parent materials are here influencing soil properties.

(g) Discuss ways in which the results of this study might be improved and the conclusions could be made more decisive.

8

Confidence Intervals and Estimation from Samples

IN THE previous chapter we considered the factors necessary to ensure that a representative and accurate sample is obtained. The present chapter is concerned with assessing the precision of a sample estimate, once it has been obtained. In other words, interest is focused on the limits, either side of a sample estimate, within which the corresponding population parameter can be inferred to lie. Here, the precision of a sample mean will be considered and the result will be expressed in the form of a confidence interval. A *confidence interval* is defined as the interval, either side of a sample estimate, within which the population parameter is expected to lie, at a known level of probability.

In Fig. 38, a population with mean (μ) and standard deviation (σ) is represented as a normal curve. Every individual in the population lies somewhere beneath the curve and there are fixed probabilities associated with particular areas (defined by the normal distribution function, Table A). Usually the mean and standard deviation of a population (the population parameters) are not known; instead, sample estimates have been obtained. For example, the population mean pebble size on a single beach would not be known but a sample mean pebble size might be available based on a large sample size of 1000 individual pebbles. How likely is it that this sample mean will correspond with the true (population) mean?

Consider Fig. 38. Any sample mean derived from this population must lie closer to the population mean (μ) than do some of the extremely large or extremely small individual values (pebbles, for example). This is because a mean evens out individual differences. Indeed, a particular sample mean must lie somewhere beneath another curve (indicated as a broken line in Fig. 38), which has the same mean as the population curve but a lower dispersion or spread. This second curve is known as a *sampling distribution*, in this case a *distribution of sample means*. This is the distribution that would result if a large number of samples were taken from the same population (each sample based on the same sample size) and the sample means were plotted in the form of a curve.

It is known, therefore, that a particular sample mean lies somewhere beneath the broken curve in Fig. 38. The accuracy of this sample mean depends on how close it is to the population mean. Given that the distribution of sample means is normally distributed, then the probability of a particular sample mean being any number of *standard deviations of the distribution of sample means* from the population mean can be found. Note that we

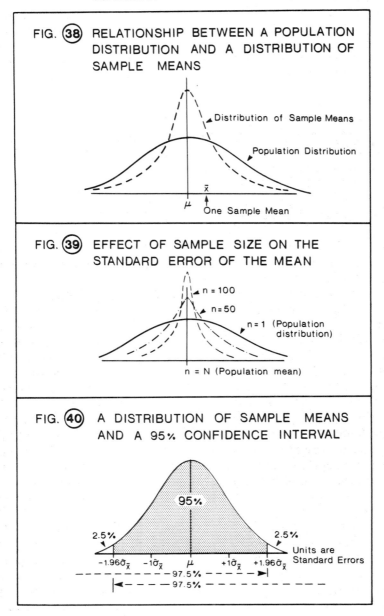

FIG. (38) RELATIONSHIP BETWEEN A POPULATION DISTRIBUTION AND A DISTRIBUTION OF SAMPLE MEANS

Distribution of Sample Means

Population Distribution

$\bar{x}$

μ

One Sample Mean

FIG. (39) EFFECT OF SAMPLE SIZE ON THE STANDARD ERROR OF THE MEAN

n = 100

n = 50

n = 1 (Population distribution)

n = N (Population mean)

FIG. (40) A DISTRIBUTION OF SAMPLE MEANS AND A 95% CONFIDENCE INTERVAL

95%

2.5%

2.5%

Units are Standard Errors

$-1.96\hat{\sigma}_{\bar{x}}$ $-1\hat{\sigma}_{\bar{x}}$ μ $+1\hat{\sigma}_{\bar{x}}$ $+1.96\hat{\sigma}_{\bar{x}}$

97.5%

97.5%

are not concerned here with the sample standard deviation but with the standard deviation of the broken curve, known for short as the *standard error of the mean*. The standard error of the mean is easily estimated from the sample standard deviation and the sample size, because it is determined by the population standard deviation and the sample size. If the population is characterized by great variability and/or if the sample size is small, then the standard error of the mean will be large. The effect of sample size on the standard error of the mean is portrayed in Fig. 39. As sample size is increased, so the spread or dispersion of the sampling distribution (the distribution of sample means) is reduced—the result of the relative smoothing power of a mean based on a large sample size. Note that as sample size is reduced, so the distribution of sample means approximates to the population distribution (identical to a sampling distribution with $n = 1$).

The standard error of the mean is, in fact, proportional to the population standard deviation (σ) and inversely proportional to the square root of sample size:

$$\sigma_{\bar{x}} = \frac{\sigma}{\sqrt{n}} = \text{standard error of the mean (symbol is sigma subscript } \bar{x}).$$

When the population standard deviation is unknown (almost always) the best estimate of the standard error of the mean is calculated by:

$$\hat{\sigma}_{\bar{x}} = \frac{s}{\sqrt{n-1}} = \frac{\hat{\sigma}}{\sqrt{n}} = \text{best estimate of the standard error of the mean}$$

where s = the sample standard deviation,
$\hat{\sigma}$ = the best estimate of the population standard deviation (derived from s with the application of Bessel's correction).

We can now place *confidence intervals* around a sample mean. Because one sample mean varies from the next to some unknown degree, only probability statements can be made about the value of the true or population mean; definite, absolute values cannot be given. However, an interval, within which the population mean is likely to lie at a selected level of probability can be defined precisely. For example, it is approximately 68 % certain that the true population mean (μ) lies between ± 1 standard error of the mean from the sample mean, that is:

$$\bar{x} \pm 1.0\,\hat{\sigma}_{\bar{x}} \text{ (probability level approximately 68 \%)}$$

If we want to be more certain of the limits within which the population mean is likely to be found, then a higher probability level must be used, which results in a broader confidence interval. The 'rule of thumb' for 95 % certainly is to use a confidence interval of ± 2 standard errors of the mean:

$$\bar{x} \pm 2.0\,\hat{\sigma}_{\bar{x}} \text{ (probability level approximately 95 \%)}$$

The 95 % probability level is the most commonly used in Geography. That is, the population parameter is expected to lie within the given intervals on 95 % of the occasions on which it is used; the corollary being that on one occasion in twenty (5 % of occasions) the true mean will lie outside the set limits.

Confidence intervals for particular levels of probability can be found by employing the normal distribution function (Table A). In general:

$$\bar{x} \pm z \cdot \hat{\sigma}_{\bar{x}} \text{ (probability level determined by } z)$$

The table shows that a 95 % confidence interval is set, strictly speaking, at ± 1.96 standard errors of the mean, not at 2.0 standard errors of the mean. Figure 40 illustrates how this value for z is found. A diagram is always useful to avoid obtaining the wrong area under the normal curve. It should be noted that there is considerable similarity between the use of the normal distribution function for setting up confidence intervals *around a sample mean* and its use in Chapter 4 for making probability statements *about individual measurements.* Normal distributions are involved in both applications, but the setting up of confidence intervals involves a sampling distribution (the distribution of sample means) whereas in the previous chapter the distribution of individual measurements was involved.

It has already been pointed out that a distribution of sample means can be envisaged as the normal distribution that would result from the plotting of a very large number of sample means, which had been drawn from the same population. This is shown in Fig. 41, in which twelve sample means are represented and the corresponding sampling distribution shown in relation to the fixed value of the population mean (μ). This figure clarifies why it is that a 95 % confidence interval placed around a sample mean is 95 % certain to include the true mean value. Because the sampling distribution is normally distributed, there is a 95 % probability that any particular sample mean will lie within ± 1.96 standard errors of the mean from the population mean. The 95 % confidence interval that has been placed around each sample mean (the horizontal bars in Fig. 41) has a width of ± 1.96 standard errors of the mean. Only when a sample mean lies to the left of

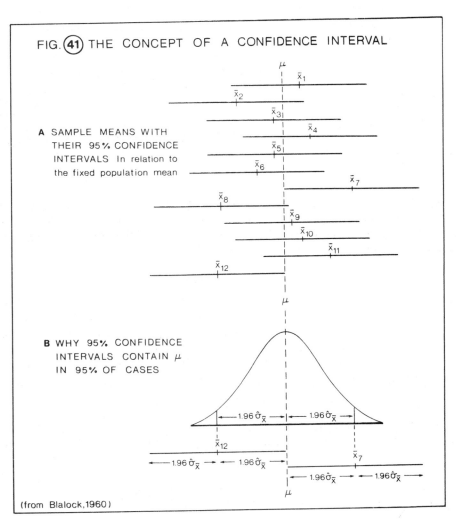

FIG. (41) THE CONCEPT OF A CONFIDENCE INTERVAL

A SAMPLE MEANS WITH
  THEIR 95% CONFIDENCE
  INTERVALS In relation to
  the fixed population mean

B WHY 95% CONFIDENCE
  INTERVALS CONTAIN μ
  IN 95% OF CASES

(from Blalock, 1960)

$\bar{x}_{12}$ or to the right of $\bar{x}_7$ will the confidence interval fail to include μ, and this is likely to happen on only 5 % of occasions. One's faith, therefore, should be placed in the approach, rather than the precise limits of a particular confidence interval.

Calculation of confidence intervals using Student's t

The normal distribution function is appropriate for the calculation of confidence intervals only when the sample size is large. When sample size is small, z is replaced by t. *Student's t tables* are used in much the same way as tables of the normal distribution function (Table C, Appendix), although they are arranged in a more convenient way for the calculation of confidence intervals. In the table, t values are given corresponding to a particular probability level (across the top of the table) and to particular degrees of freedom (down the left-hand margin). *Degrees of freedom* $(v = n - 1)$ is one less than the sample size; the probability value given (p) is the *significance level*. For example, to set a 95 % confidence interval around a sample mean based on a sample size of 20, $p = 5$ % and $v = 19$; the t value is thus 2.093. It can be seen that as sample size increases so the t value approaches the corresponding value of z (1.96 at the 5 % significance level). The value of t differs markedly from z at very small sample sizes, which results in a wider confidence interval when t is used. In other words, use of z at small sample sizes gives rise to a false sense of precision.

The formula for a confidence interval about a sample mean can therefore be rewritten as:

$$\bar{x} \pm t \cdot \hat{\sigma}_{\bar{x}} \quad \text{(probability level determined by } t)$$

or

$$\bar{x} \pm t \cdot \frac{s}{\sqrt{n-1}}$$

The necessary calculations in relation to actual data are given in the following worked example.

Lichenometric dating was applied to the outermost end moraine in front of the Storbreen glacier in the Jotunheimen Mountains of southern Norway. From lichen measurements made in A.D. 1974, the following mean age of the moraine was obtained, with a sample size of 100 (100 datings were obtained):

Data	$\bar{x}$	s
Raw data	229 years	75 years
Log-transformed data	2.336 46	0.1306

Transformed data were used to account for a positive skew shown by the raw data (Fig. 42). The problem is to estimate the confidence interval within which the true age of the moraine lies at a selected probability level. The stages involved in the calculation are:
(a) Select a probability level. Here the 95 % confidence interval will be used.
(b) Calculate the best estimate of the standard error of the mean $(\hat{\sigma}_{\bar{x}})$, using the sample standard deviation (s) and the sample size (n).

$$\hat{\sigma}_{\bar{x}} = \frac{s}{\sqrt{n-1}} = \frac{0.1306}{\sqrt{99}} = 0.0131.$$

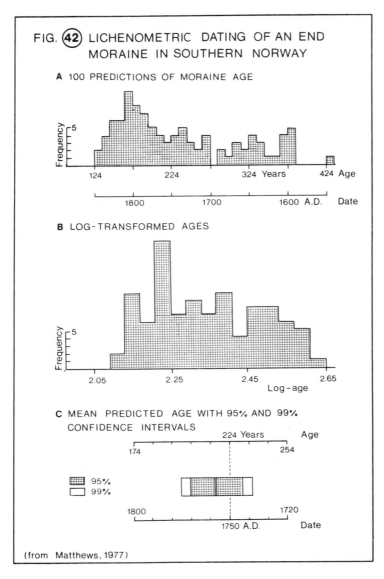

FIG. (42) LICHENOMETRIC DATING OF AN END
MORAINE IN SOUTHERN NORWAY

A 100 PREDICTIONS OF MORAINE AGE

Frequency

124   224   324 Years   424 Age

1800   1700   1600 A.D.   Date

B LOG-TRANSFORMED AGES

Frequency

2.05   2.25   2.45   2.65
Log-age

C MEAN PREDICTED AGE WITH 95% AND 99%
CONFIDENCE INTERVALS

224 Years   Age
174   254

95%
99%

1800   1720
1750 A.D.   Date

(from Matthews, 1977)

(c) Obtain the appropriate t value from Table C (Appendix); using 99 degrees of freedom $(n-1)$ and a 5% significance level for the 95% confidence interval, $t = 1.984$.

(d) The required confidence interval is therefore 1.984 standard errors of the mean (around the sample mean), or

$$\bar{x} \pm t \cdot \hat{\sigma}_{\bar{x}} = 2.336\,46 \pm 1.984\,(0.0131) = 2.336\,46 \pm 0.0259$$

(e) In other words, we can be 95% certain that the true mean age of the moraine lies between antilog 2.3105 and antilog 2.3623 years, or between 204 and 230 years. That is, the moraine dates from between A.D. 1744 and A.D. 1770 with 95% certainty.

This result is shown graphically in Fig. 42C. If the chance of failing to include the true mean within the confidence interval is to be minimized even further, then a 99 % confidence interval might be appropriate. This would, of course, result in a broader interval, with greater certainty gained at the expense of precision. Conversely, a narrower confidence interval around the sample mean is a more precise statement about the value of the true mean, but it is less certain to enclose the true mean. The more precise the estimate, the lower the confidence that can be placed in it.

In conclusion it can be stated that confidence intervals are controlled by three factors: (1) the variability of the data, reflected by the standard deviation of the sample and dependent on the variability of the underlying population; (2) the size of the sample on which the sample estimate is based; and (3) the probability level or certainty with which one wishes to express conclusions. All three factors are represented in the above formula for a confidence interval, which will give meaningful results provided that the sample is unbiased and provided that the individual measurements comprising the sample are statistically independent.

Exercise 11. Use of tables of the t distribution and the calculation of confidence intervals around sample means.

Background

This exercise concentrates on the manipulation of tables of Student's t (Table C, Appendix) and the calculation of confidence intervals around sample means; means that have been calculated in previous exercises. Particular attention should be paid to the different concepts behind the present exercise compared to the concepts underlying Exercise 6.

Practical work

1. (a) Using Table C, what number of standard errors of the mean, placed either side of a sample mean, is appropriate to define confidence intervals with the following probability levels and sample sizes:

Probability level	Sample size			
	10	25	60	∞
90 %				
95 %				
99 %				

(b) Explain, with the aid of diagrams to represent areas under a normal curve, how the values of t in the last column of the above question can be obtained from Table A (the normal distribution function).

(c) Based on a sample size of thirty individual measurements, the following confidence intervals have been placed around a sample mean:

(i) $\pm 1.70\,\hat{\sigma}_{\bar{x}}$

(ii) $\pm 2.04\,\hat{\sigma}_{\bar{x}}$

(iii) $\pm 2.75\,\hat{\sigma}_{\bar{x}}$

For each of these confidence intervals:

(i) How certain are you that the true mean lies within the interval?

(ii) What is the probability that the true mean is not enclosed within the interval?

(iii) What is the probability that the true mean lies above the upper confidence limit?

(iv) What is the probability that the true mean lies above the lower confidence limit *but* has a lower value than the sample mean?

(d) Based on a sample size of 28, at what number of standard errors of the mean must confidence limits be set to ensure the following:

(i) That there is a less than 1 in 10 chance of the true mean lying outside the confidence interval.

(ii) That there is at most a 5 % chance that the true mean lies below the lower confidence limit.

(iii) That you are at least 45 % certain that the true mean lies between the sample mean and the upper confidence limit.

2. Using your answers to question 1, Exercise 5, relating to the mean and standard deviation of June and December monthly rainfall totals at Cardiff over a 22-year period, calculate the following:

(a) The interval within which the true mean rainfall for June is expected to lie with 95 % certainty. The true mean can be envisaged as the mean for an extremely long run of years.

(b) The maximum and minimum quantities of rain between which the true mean rainfall for December is expected to lie with at least a 95 % probability of being correct.

(c) What can be concluded from your answers to 2 (a) and 2 (b) regarding the difference in raininess of June and December at this station?

(d) Set 90 % confidence intervals around the June and December means and explain the implications of the change in width of the interval for 2 (c).

(e) Would a longer period of record result in a wider or a narrower confidence interval at a particular probability level? Explain your answer.

(f) What is the probability that the true mean rainfall for June lies within $\pm 25\,mm$ of the sample mean?

(g) What is the probability that the true mean rainfall for June lies more than 25-mm above the sample mean?

3. The following means and standard deviations relate to the data on erratic size (cm^2) given in Exercise 4, question 1:

Descriptive statistic	Distance from the outcrop (km)									
	1	3	5	7	9	11	13	15	17	19
Sample mean $(\bar{x})$	491	350	412	435	469	363	407	286	255	316
Standard deviation $(\hat{\sigma})$	374	292	196	208	256	212	309	210	135	342

(a) Calculate the 95 % confidence interval for each sample mean and represent the results in a graph of erratic size plotted against distance from the outcrop.

(b) In the light of the confidence intervals, comment on the patterns shown by your graph, and the adequacy of the sample size used in this investigation.

(c) Using the results from the 1-km distance, what sample size would be necessary to reduce the width of the confidence interval to at most $\pm 50 \text{ cm}^2$?

Exercise 12:   Use of confidence intervals in the evaluation of models of vegetation succession in the Jotunheimen Mountains of southern Norway.

Background

The sequence of vegetation that invades a newly formed habitat, follows forest clearance, or is initiated by the abandonment of agricultural land, is known as a vegetation succession. Vegetation successions are in progress almost everywhere, but little precise information is available concerning such changes, except over rather short time-spans. In only a limited number of situations is it possible to gain detailed information about longer-term successional change. The present exercise is based on one such situation, where progressively older terrain is found with increasing distance from the margin of a retreating glacier.

Figure 43A shows the area in front of Storbreen, Jotunheimen, southern Norway, where the course of deglaciation has been reconstructed in detail from historical and lichenometric evidence. On this terrain of known age, the plant species have been recorded at each of 638 sites (each site being an area of 16 m² located according to a systematically stratified random sampling design).

From the study of vegetation on ground of increasing age, the progress of a vegetation succession can be traced. For example, Figs. 43 B-D depict the colonization patterns of three plant species, and indicate that the Tufted saxifrage (*Saxifraga groenlandica*) is replaced by the Mountain sorrel (*Oxyria digyna*), which is in turn replaced by the Arctic crowberry (*Empetrum hermaphroditum*) on progressively older terrain. However, it is not the intention here to deal with individual species, but to deal with the total number of species at each site.

There are at least three alternative views about the changes that occur in the number of species during a vegetation succession (Fig. 44):

(1) There is a steady increase in numbers towards an equilibrium state.

(2) There is a rapid increase in numbers followed by an observable decrease to an equilibrium state, as well adapted species become dominant.

(3) There is a rapid increase in numbers, followed by a decrease, followed by further waves of immigration and a succession of dominant species.

The purpose of this exercise is to evaluate these alternatives in the light of data from in front of Storbreen (Table 8). The number of species per site is given in ten zones of increasing age, which permits the calculation of the mean number of species per site for each zone, and hence the construction of a graph of the type shown in Fig. 44. Use of confidence intervals will be found crucial for deciding on the degree, direction and number of meaningful changes from zone to zone.

TABLE 8. *The number of species on terrain of increasing age in front of Storbreen, southern Norway*

Age of zone (mid-point in years before 1970)									
5	15	25	32	48	81	109	139	190	> 220
4	10	12	13	15	9	20	19	10	17
2	8	23	17	8	12	15	19	18	18
6	11	17	27	15	11	16	23	30	23
6	11	20	24	11	8	19	17	21	28
6	13	13	22	24	13	14	13	27	29
5	10	18	29	15	18	15	14	24	26
7	4	11	26	5	16	20	18	28	30
6	10	13	10	19	11	12	24	27	13
6	13	20	15	13	17	15	16	32	20
6	16	18	15	12	11	19	12	21	13
6	12	18	24	18	18	16	14	16	14
8	10	24	10	10	11	9	10	17	15
9	15	14	14	10	17	14	17	25	9
4	15	19	9	14	13	14	20	23	13
6	12	8	10	11	10	14	20	22	16
7	12	13	16	13	12	7	18	28	18
	17	13	19	13	23	16	23	28	28
	14	16	22	13	20	15	22	29	28
	14	13	24	12	11	15	10	17	19
	7	7	11	18	6	13	8	25	24
	7	7	13	10	12	20	5	23	16
	13	12	12	12	11	13	11	13	11
	17	9	16	20	20	13	10	26	17
	17	13	20	16	10	9	20	14	12
	18	10	16	12	9	12	9	9	21
	11	18	22	16	13	14	11	9	13
	24	14	10	12		11	10	9	19
	16	11	15	10		13	11	6	19
	8	13		13		16		15	23
	19	11		13		11		14	14
		10		21		13		10	12
		6		14		12		10	17
				10		14			10
				16		12			23
				16		17			14
				21		9			14
				17		25			14
				14		10			23
				10		14			7
				12					13
				7					14
				23					
				15					
				18					
				18					
				15					
				14					
				23					
				15					
				21					
				16					

Each value is the number of
species at a 16 m² site

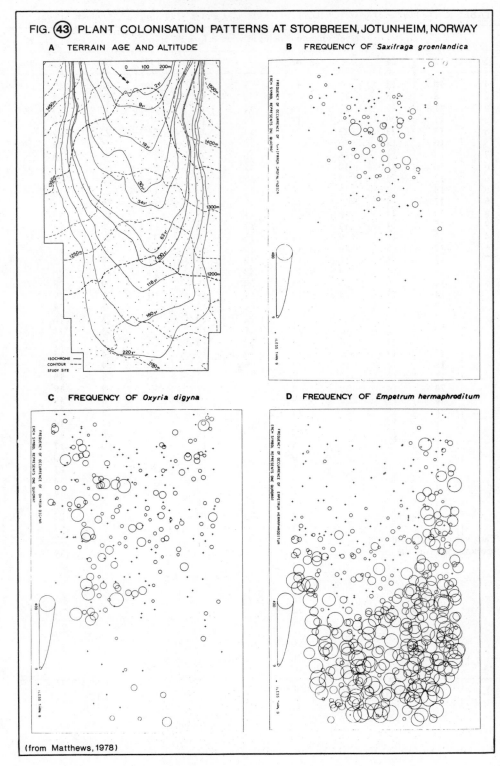

FIG. ㊸ PLANT COLONISATION PATTERNS AT STORBREEN, JOTUNHEIM, NORWAY

A   TERRAIN AGE AND ALTITUDE

B   FREQUENCY OF *Saxifraga groenlandica*

C   FREQUENCY OF *Oxyria digyna*

D   FREQUENCY OF *Empetrum hermaphroditum*

(from Matthews, 1978)

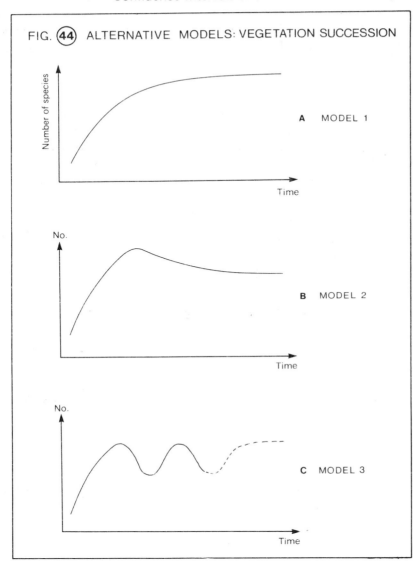

FIG. (44) ALTERNATIVE MODELS: VEGETATION SUCCESSION

A   MODEL 1

B   MODEL 2

C   MODEL 3

Practical work

1. For each time zone, calculate the following:
(a) The mean number of species per site.
(b) The sample standard deviation.
(c) The standard error of the mean.
(d) The 95 % confidence interval about the sample mean.

Summarize your results in the form of a table:

	5	15	25	32	48	..., etc.	Zone age
$\bar{x}$							
s							
$\hat{\sigma}_{\bar{x}}$							
$t \cdot \hat{\sigma}_{\bar{x}}$							

2. Plot a graph showing the relationship between mean number of species and terrain age. Your graph should depict clearly the mean number of species for each zone, together with its confidence interval. The graph should be drawn in such a way that it is comparable to the graphs in Fig. 44.

3. Describe the graph, paying particular attention to:

(a) Its overall form.

(b) Any significant 'wiggles'.

4. Discuss the probable effect on your answer to 3(b) if (i) 90 % or (ii) 99 % confidence intervals had been used.

5. Evaluate the models of vegetation succession (Fig. 44), in the light of the evidence from in front of Storbreen, paying particular attention to:

(a) The 'best' model (if any).

(b) Uncertainties remaining as to the precise form of the succession at Storbreen.

6. Suggest some possible explanations for the form of the vegetation succession at Storbreen.

9

Statistical Hypothesis Testing Based on Student's *t*

Tests involving one sample

A confidence interval about a sample mean (defined and explained in Chapter 8) is the interval within which the underlying population mean is expected to be found, at a known probability level. This concept of a confidence interval can be used as the basis for the statistical testing of hypotheses relating to the population from which the sample was drawn. There may be, for example, independent evidence that suggests a *hypothesized true mean* for the underlying population. Given an unbiased sample, then it is possible to calculate the likelihood that the sample mean was drawn from the same population as the hypothesized true mean, thereby testing the *hypothesis of 'no difference'* between the sample mean and the hypothesized true mean. If the confidence interval around the sample mean fails to enclose the hypothesized true mean it must be concluded that there is a significant difference between the sample mean and the hypothesized true mean. In other words, the hypothesis of 'no difference' is rejected at a known probability level. If, on the other hand, the confidence interval around the sample mean encloses the hypothesized true mean, then the hypothesis of 'no difference' cannot be rejected. In this instance, the sample mean is more likely to have been drawn from the same population as the hypothesized true mean. By rejecting or failing to reject hypotheses of 'no difference' in this way, *statistical tests* of hypotheses are carried out at known probability levels.

The confidence interval in Fig. 42 (C) can be used to test a hypothesis in the manner outlined above. The figure shows that the true age of the outermost end moraine in front of the Storbreen glacier is expected to lie between 204 years and 230 years at the 95% probability level. These ages correspond to dates of A.D. 1744 and A.D. 1770. A hypothesized true age is suggested by documentary evidence from a neighbouring glacier (Nigardsbreen), which attained its historical maximum extent about 1750. We are now in a position to test the hypothesis that there is no difference between the estimated age of the Storbreen end moraine and the known age of the Nigardsbreen moraine. Because there is a 95% probability that the true date of the Storbreen moraine lies between A.D. 1744 and A.D. 1770, we cannot reject the possibility of the moraine dating from A.D. 1750. The phrasing of the preceding sentence is very important; it is *not* valid, for example, to state that an A.D. 1750 date is accepted, only that this date cannot be rejected, for there are a large number of alternative dates within the confidence interval. Similarly, it is *not* valid to state that the hypothesis of 'no difference' is accepted, only that the difference is not sufficiently great to permit rejection (at the 95% confidence level).

This example has shown how confidence intervals may be used to test hypotheses. The same test of the difference between a sample mean ($\bar{x}$) and a hypothesized true mean (μ) may be expressed more formally as a *one-sample Student's t-test*, which has three steps:

(a) Calculation of the difference between the sample mean and the hypothesized true mean in terms of standard errors of the mean. This will be termed the *calculated t statistic*.

(b) Assessment (by reference to tables of Student's t, Table C, Appendix) of the difference that would be expected to occur by chance (as a result of sampling from a population with the hypothesized true mean). This will be termed the *tabulated t statistic*.

(c) Comparison of the calculated and tabulated t values. If the calculated t exceeds the value that is likely to result by chance, then the hypothesis of 'no difference' between the sample mean and the hypothesized true mean can be rejected, and the difference is said to be statistically significant.

Application of these three steps to the above example gives the following results (using log-transformed values throughout).

(a) The calculated t becomes:

$$t = \frac{(\bar{x} - \mu)}{\hat{\sigma}_{\bar{x}}} = \frac{\text{Difference between sample mean and hypothesized true mean}}{\text{Standard error of the mean}}$$

$$= \frac{2.336\,46 - 2.350\,25}{0.1306/\sqrt{99}} = 1.05$$

(b) The tabulated t depends on the sample size and the probability level and is derived from Table C. With a sample size of 100 (that is, $n - 1 = 99$ degrees of freedom) and a 95% confidence level (that is, a 5% significance level), a t value as high as 1.98 is expected to occur by chance if the true mean were indeed 224 years. This value for the tabulated t is illustrated in Fig. 45, which shows that a difference in excess of $t = \pm 1.98$ is likely to occur with a probability of less than 5% (the critical shaded area in the figure).

(c) Because the calculated t statistic does not exceed the value of t that is likely to occur by chance, we are unable to reject the hypothesis of 'no difference' from a true age of 224 years. The sample mean does not, therefore, differ significantly from the hypothesized age at the 5% level of significance (the 95% confidence level).

This means that there is a less than 95% probability of an actual difference. In other words, if the hypothesis of 'no difference' were to be rejected, then there would be a greater than 5% chance of having made the wrong decision. In fact, the calculated t also fails to exceed the value of t that has a probability of occurrence by chance of 10% (see Table C); thus it is possible to say that it is less than 90% likely that the sample mean differs significantly from a true age of 224 years.

Two-sample difference of means test

In the section above, a hypothesis relating to a single sample mean ($\bar{x}$) and a hypothesized true mean (μ) was tested. A modification of the one-sample Student's t-test

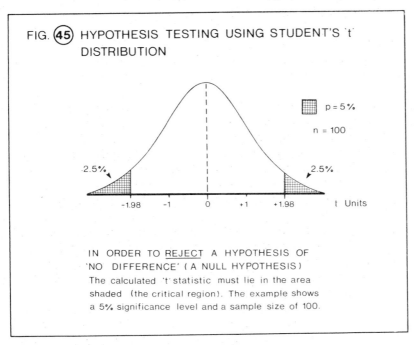

FIG. ④⑤ HYPOTHESIS TESTING USING STUDENT'S 't'
DISTRIBUTION

p = 5%

n = 100

2.5%

2.5%

-1.98   -1   0   +1   +1.98   t Units

IN ORDER TO REJECT A HYPOTHESIS OF
'NO DIFFERENCE' (A NULL HYPOTHESIS)
The calculated 't' statistic must lie in the area
shaded (the critical region). The example shows
a 5% significance level and a sample size of 100.

gives rise to the *two-sample Student's t-test* of a difference between two sample means ($\bar{x}_1$ and $\bar{x}_2$). In the one-sample test, the standard error of the mean was used; in the two-sample test, we use the standard error of the difference between two means or, for short, the *standard error of the difference*. The standard error of the mean is given by:

$$\hat{\sigma}_{\bar{x}} = \frac{s}{\sqrt{n-1}}$$

The standard error of the difference is given by:

$$\hat{\sigma}_{\bar{x}_1 - \bar{x}_2} = \sqrt{(\hat{\sigma}_{\bar{x}_1})^2 + (\hat{\sigma}_{\bar{x}_2})^2} = \sqrt{\left(\frac{s_1}{\sqrt{n_1 - 1}}\right)^2 + \left(\frac{s_2}{\sqrt{n_2 - 1}}\right)^2}$$

Recollect that in the one-sample test we calculate the number of standard errors of the mean that a sample mean lies from a hypothesized true mean (the calculated *t* statistic); in the two-sample test interest focuses on the number of standard errors of the difference that one sample mean lies from the other sample mean. If the difference between the two means is greater than the appropriate tabulated *t* statistic, then the difference between the two means lies in the critical shaded region of Fig. 45, and the difference is greater than would be expected to occur by chance between two sample means drawn from the same population. If the calculated *t* is greater than the tabulated *t*, then the hypothesis of 'no difference' between the two sample means is rejected, at a particular significance level.

For the two-sample test the calculated t statistic is found from:

$$t = \frac{\bar{x}_1 - \bar{x}_2}{(\hat{\sigma}_{\bar{x}_1 - \bar{x}_2})} = \frac{\text{Difference between two sample means}}{\text{Standard error of the difference}}$$

The appropriate tabulated t statistic is obtained from Table C in the usual way, using $(n_1 - 1) + (n_2 - 1)$ degrees of freedom, where n_1 and n_2 are the two sample sizes.

An example is provided by the results of an experiment on the growth of the southern beech (*Nothofagus*) near the timber line in New Zealand (Wardle, 1971). Wardle found that seedlings grown at an altitude of 1100 m and then transplanted to 1600 m grew at a mean rate of 4.9 mm/week (based on a sample size of 15 seedlings), whereas seedlings grown at 1100 m and then transplanted to another site at 1100 m grew at a mean rate of 49.1 mm/week (based on a sample size of 11 seedlings). Given that the standard errors of the two means were 0.8 mm/week and 7.5 mm/week, respectively, is there a significant difference between the two sample means at the 5% significance level?

Following the same three steps defined previously in relation to a one-sample test, the appropriate two-sample test can be outlined.

(a) The calculated t statistic is

$$t = \frac{\bar{x}_1 - \bar{x}_2}{\sqrt{(\hat{\sigma}_{\bar{x}_1})^2 + (\hat{\sigma}_{\bar{x}_2})^2}} = \frac{4.9 - 49.1}{\sqrt{0.8^2 + 7.5^2}} = \frac{-44.2}{7.5425} = -5.86$$

(b) The tabulated t statistic is derived from Table C using the 5% significance level, and $(n_1 - 1) + (n_2 - 1) = 14 + 10 = 24$ degrees of freedom. Thus a t value as high as 2.06 would be expected if the two means were drawn from the same population.

(c) The calculated t exceeds the tabulated t. It is therefore concluded that the difference between the sample means is greater than is likely to occur by chance at the 5% significance level. In other words, there is a less than 5% probability of making a wrong decision by rejecting the hypothesis of 'no difference' between the two means. Similarly, one is greater than 95% certain that the two means reflect a real difference.

Wardle carried out a second experiment in which seedlings were germinated and grown at 1600 m and at 1100 m. He found that the two sample means were not significantly different. The results of his two experiments suggest that the growth of the southern beech near the timber line is strongly influenced by the development of 'hardiness' in the early stages of growth. This hardiness was not developed by the seedlings that were germinated at 1100 m and transplanted to 1600 m. From these and similar experiments involving statistical testing of the differences between sample means, Wardle went on to propose an explanation for alpine timberlines in general.

Tests involving 'dependent' (matched) samples

The two-sample Student's t-test, like most statistical tests used in Geography, assumes *independent sampling*. That is, each individual in each sample must be selected independently of each other. This condition might be achieved by drawing a random sample from the first population, followed by a random sample from the second population (without any reference to the first). However, certain types of *controlled dependence* can be an advantage in the context of hypothesis testing, provided that an appropriate modified test is used. The comparison of 'before' and 'after' situations, when a group of people is interviewed twice, provides a good example of where such a test is appropriate. The interviewees might be selected at random, but if the same group is interviewed after a set period, then the second sample is not independent of the first. Nevertheless, this type of data is extremely valuable, because many 'interfering' variables are controlled or held constant if the matched pairs of answers are analysed.

An appropriate test for this kind of matched data is the *Student's t-test of a difference between two dependent means*. The calculated t statistic is obtained from the pair by pair differences and involves the initial calculation of the *mean difference* ($\bar{x}_D$) and the *standard deviation of the differences* (s_D):

$$t = \frac{\bar{x}_D}{\frac{s_D}{\sqrt{n-1}}} = \frac{\text{Mean difference between matched pairs off individuals}}{\text{Standard error of the mean difference between matched pairs}}$$

The degrees of freedom for the test are $(n-1)$ and the tabulated t statistic is obtained in the same way as in a two-sample test of the difference between independent means.

Schumm's (1956) study of badland slopes at Perth Amboy, New Jersey, which was referred to in Chapter 2 (Fig. 10), will be used as a worked example of the application of the test. Slope angles were measured in 1949 at 149 points; the same points were remeasured in 1952. The mean difference in slope angle was found to be $-0.21°$; in other words the slopes had, on average, declined by $0.21°$. The standard deviation of the differences was found to be $3.76°$. Is the apparent decline in slope angle statistically significant at the 5 % level?

The three steps for the test are as follows:
(a) The calculated t statistic is

$$t = \frac{\bar{x}_D}{\frac{s_D}{\sqrt{n-1}}} = \frac{-0.21}{\frac{3.76}{\sqrt{148}}} = \frac{-0.21}{0.309\,07} = -0.68$$

(b) The tabulated t statistic is derived from Table C using the 5 % significance level and $(n-1) = 148$ degrees of freedom. Thus a t-value of up to 1.97 would be expected to be the result of chance.

(c) The calculated t (the measured difference in terms of standard error units) does not exceed the tabulated t. It can therefore be concluded that the measured difference is less than is likely to occur by chance at the 5 % significance level, so that the hypothesis of 'no difference' cannot be rejected. In other words, there is a greater than 5 % probability of making a mistake if the hypotheses of 'no difference' is rejected, and we are less than 95 % certain of a difference.

It is interesting to note that Schumm (1956) concluded from this test that his measurements indicated 'uniform lowering' of his slopes. He claimed these results to be evidence against the concept of slope decline through time. These conclusions were not entirely justified, however. There is a less than 95 % probability of a difference in slope angle over the period of interest, which is far removed from being confident of 'no difference' (necessary for 'uniform lowering' to be accepted).

The language of statistical hypothesis testing

The three Student's *t*-tests that have been outlined in this chapter are governed by the same statistical concepts and rules. The purpose of this section is to define some terms which, although not necessary for understanding the principles, are commonly used in association with any statistical test, and will be encountered in further reading on this topic.

The term *null hypothesis* (H₀) is commonly used to describe a hypothesis of 'no difference'. This is the hypothesis that is actually tested in a statistical test and should be distinguished from the alternative hypothesis (H₁). The null hypothesis is testable because it is an exact, precise statement, whereas there are many alternatives to the hypothesis of 'no difference'. In particular, a precise null hypothesis is necessary for the computation of a *sampling distribution* (such as the tabulated *t* statistic), which describes the *expected probabilities* of all outcomes of a test, assuming the null hypothesis to be true (that is, assuming there is in fact 'no difference').

Any test of a null hypothesis is carried out at a given *level of statistical significance* (α). This is the probability of making a *type 1 error*, which is defined as *rejecting a true null hypothesis*, or recognizing a difference that does not really exist. Use of the 5 % ($p = 0.05$) significance level ensures that there is only a small (1 in 20) chance of committing a type 1 error. By minimizing the probability of making a type I error, the probability of having made the correct decision (the confidence level) is maximized. However, in so doing, the probability of making a *type II error* (*failing to reject a false null hypothesis*, or failing to recognize a real difference) increases. Although it is considered that first priority should be given to the avoidance of type I errors, the sensible use of statistical tests in Geography requires that the possibility of type II error should not be ignored. This is a similar problem to the one encountered in the setting of confidence intervals; too broad a confidence interval and too stringent a significance level are related by their inability to detect differences.

When a statistical test involves an unqualified hypothesis of 'no difference', then the critical region for rejecting, or failing to reject, the null hypothesis involves both ends (tails) of the sampling distribution. All the tests described in this chapter (and illustrated in Fig. 45) have been of this type, and are known as *two-tailed tests*. It is also possible to test null hypotheses of the form 'no greater than' or 'no less than'. This kind of hypothesis involves a *one-tailed test*. In a one-tailed test, the *critical region* for rejecting the null hypothesis lies in one tail only (Fig. 46A). In practice, the tabulated *t* statistic has a lower value for a one-tailed test than for a two-tailed test. In the worked example concerning the growth of southern beech seedlings near the tree line in New Zealand, the hypothesis that growth at 1600 m was 'no different' from growth at 1100 m was tested. The tabulated *t* statistic for a 5 % significance level and 24 degrees of freedom was 2.06. A test of the hypothesis that growth at 1600 m was 'no less than' growth at 1100 m, for the same significance level and degrees of freedom, results in a tabulated *t* statistic of 1.71. Reference to Fig. 46 shows that the required tabulated *t* for a one-tailed test at the 5 % significance level is equivalent to the tabulated *t* for a two-tailed test at the 10 % significance level, because the area in the lower tail is twice as great for a one-tailed test as for a two-tailed test. Whether a one- or two-tailed test is chosen will depend on the nature of the research problem and in particular the availability of reliable independent knowledge, which would have some bearing on the formulation of hypotheses. If in doubt, however, a two-tailed test should be used.

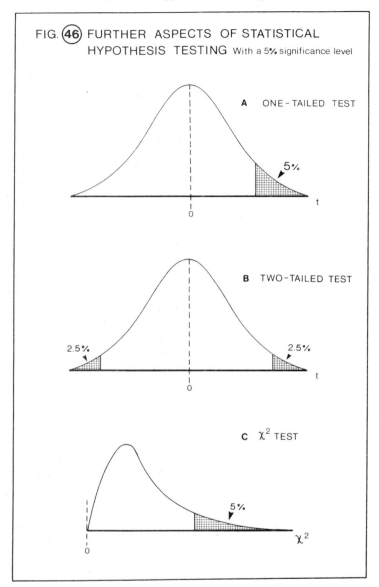

FIG. (46) FURTHER ASPECTS OF STATISTICAL HYPOTHESIS TESTING With a 5% significance level

A   ONE-TAILED TEST

5%

t

0

B   TWO-TAILED TEST

2.5%

2.5%

t

0

C χ^2 TEST

5%

χ^2

0

Exercise 13. Hypothesis testing about the upper limit of agriculture on the North York Moors [north-east England] using Student's *t*-tests.

Background

The moorland edge is an important feature of the landscape in most upland areas where the semi-natural, extensively-used wildscape impinges on the more intensively-used, man-made farmscape. Its prominence as a morphological feature, and its sensitivity to the

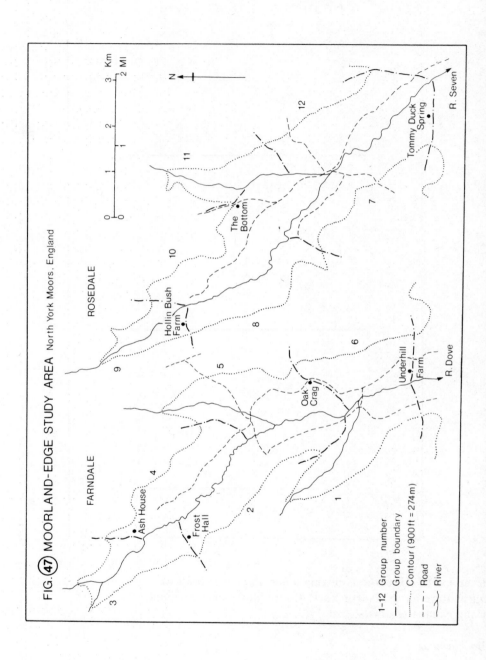

FIG. (47) MOORLAND-EDGE STUDY AREA North York Moors, England

1-12   Group number
— · —   Group boundary
....   Contour (900 ft = 274 m)
— — —   Road
⌇   River

TABLE 9. *Data relating to the moorland edge, North York Moors, north-east England*

Group no.	Altitude (ft)	Aspect	Up-slope angle	Down-slope angle	Group no.	Altitude (ft)	Aspect	Up-slope angle	Down-slope angle
FARNDALE									
1	750	E	25°	14°	5	900	S	11°	11°
1	700	E	21	11	5	910	S	14	15
1	650	NE	16	3	5	900	SW	19	13
1	750	NE	15	6	5	1000	SW	13	22
1	800	SW	21	14	5	700	W	13	11
1	600	SW	18	15	5	800	W	21	15
1	650	NE	21	19	5	810	W	23	13
1	600	E	32	5	5	810	W	23	14
1	650	NE	16	18	5	1000	WNW	25	19
2	700	E	19	12	5	825	W	18	15
2	725	E	20	13	5	900	SE	22	9
2	725	E	17	17	5	800	E	16	10
2	800	E	14	11	5	950	W	15	4
2	750	E	21	13	6	650	WNW	14	11
2	750	E	21	13	6	1000	WSW	30	21
3	No valid measurements made				6	750	WNW	25	8
4	850	WSW	15	12	6	850	W	19	12
4	675	SSE	10	9	6	750	W	30	6
4	750	SSE	10	10	6	700	NW	25	2
4	700	SSW	19	17	6	850	W	5	4
4	625	SW	14	14	6	700	W	12	12
4	750	SW	13	8	6	800	W	18	16
4	675	W	14	15	6	625	SW	19	12
4	650	W	10	8	6	650	NW	10	8
5	1000	W	15	15	6	700	W	14	12
ROSEDALE									
7	725	E	14	11	10	800	WSW	17	9
7	600	NE	17	13	10	700	S	16	8
7	625	NE	23	18	10	775	SW	18	14
7	575	NE	23	14	10	925	SW	22	17
7	650	E	18	9	10	700	SSW	16	14
7	775	E	26	19	10	800	SW	23	8
7	600	N	35	20	11	850	E	13	11
7	650	N	9	8	11	850	E	22	14
7	520	N	25	11	11	900	E	16	8
7	600	N	21	32	11	700	W	18	11
7	570	N	36	14	11	800	W	22	19
8	750	NE	20	8	11	800	W	18	11
8	750	NE	30	15	11	750	W	16	6
8	740	NE	15	10	11	850	E	12	10
8	750	SE	30	22	11	900	E	14	10
8	720	NE	20	10	11	800	W	21	17
8	730	NE	25	10	11	750	W	17	12
8	740	NE	30	15	12	875	SW	11	9
8	750	SE	30	25	12	700	SW	16	14
9	900	SW	15	15	12	700	SW	16	11
9	825	E	15	10	12	780	SW	10	10
9	900	WSW	12	10	12	850	SW	18	9
9	775	E	10	5	12	925	SW	0	3
9	775	E	15	8	12	650	SW	9	11
9	975	E	18	16	12	750	SW	9	2
9	850	W	21	13	12	600	SW	13	11
9	825	NE	30	22	12	850	SW	9	12
9	825	ESE	12	15	12	675	W	18	15
10	750	SSW	16	11	12	675	SW	8	8
10	750	SSW	17	9	12	800	SW	16	8
10	875	S	21	21	12	700	SW	10	9

present and changing physical and human factors that control its position, make it a particularly relevant topic for geographical inquiry.

Variation in the altitude of the moorland edge/agricultural limit exists from place to place, partly as a result of chance and our imperfect ability to describe and measure it, partly due to spatial variation in factors of the physical and human environment. Furthermore, the moorland edge is not a static feature; any present-day 'snapshot' reveals a feature that may or may not be in equilibrium with present conditions; the edge may be advancing, static or retreating in different locations. Above a critical altitude, at any one time and in any one place, physical factors, such as climate and slope, may limit the crops that can be grown or other uses to which the land can be put. On the other hand, the controls are not necessarily physical, for economic, social, political and other constraints may determine what a farmer does with his land.

This exercise is based on a field study undertaken by students in Rosedale and Farndale, two valleys within the North York Moors National Park (Fig. 47). Both valleys have been cut deeply into the surface of the surrounding High Moors. Farmland occupies the valley bottoms and extends for some distance up the valley sides. Geologically, the North York Moors consists of Oolitic Limestone, which dips gently to the south; in the valley bottoms fertile inliers of the underlying Liassic rocks are exposed. The object of the exercise is to test some preliminary hypotheses concerning the factors controlling the altitude of the moorland edge by judicious use of Student's t-tests.

Practical work

The data in Table 9 were collected from the moorland edge of Rosedale and Farndale. Group numbers refer to the areas located in Fig. 47. Measurements were made at equally-spaced intervals along the moorland edge in the field (a systematic sample). Altitude and aspect were found from the Ordnance Survey, 1:25,000 topographic map with a 25-feet contour interval (hence altitudes are given in feet). Slope angles were measured over a distance of 5 m either side of the moorland edge, and at right angles to the contours.

1. Using the data from either Farndale *or* Rosedale, the aim is to use a Student's t-test of the difference in altitude of the moorland edge on valley sides with different aspects.
   (a) Draw two histograms summarizing the altitude of the moorland edge on (i) north and east-facing slopes and (ii) south- and west-facing slopes (two histograms in all).
   (b) Calculate the mean and standard deviation of the altitudes on (i) north and east-facing slopes and (ii) south- and west-facing slopes.
   (c) Using an appropriate significance level, test the hypothesis that there is 'no difference' between the mean altitudes of the moorland edge on these two aspects.
   (d) Fully explain the statistical implications of the test.
   (e) Suggest some reasons for any differences detected.
2. (a) Using data for the same dale as in question 1, use a Student's t-test to determine whether or not there is a significant difference between the altitude of the moorland edge and the 900-ft contour. Separate tests should be carried out for:
       (i) North- and east-facing slopes
       (ii) South- and west-facing slopes
       (iii) All slopes.
   (b) For each test in 2 (a), what is the probability of a real difference existing between the altitude of the moorland edge and the 900-ft contour?

(c) What are the possible conclusions if a hypothesis is:
  (i) Rejected for one aspect but not rejected for the other?
  (ii) Rejected for the whole dale but not for individual aspects?
  (iii) Rejected for individual aspects but not for the whole dale?

3. Again for the same dale used in questions 1 and 2, the aim is to test the significance of the difference between slope angles found either side of the moorland edge. If slope angle is an important factor in accounting for the location of the moorland edge, then a difference in slope angle would be expected. It should be noted that the up-slope and down-slope angles are not independent samples, because the data are in the form of matched-pairs of angles (an up-slope and a down-slope angle is available for each point at which measurements were made).

  (a) Calculate the mean difference ($\bar{x}_D$) and the standard deviation of the differences (s_D) for the dale as a whole.
  (b) Carry out an appropriate Student's t-test of the difference between the up-slope and down-slope angles.
  (c) What is the probability of being wrong if the hypothesis of 'no difference' was to be rejected in this case?
  (d) Discuss the limitations of the test, paying particular attention to the ways in which slope angle is likely to influence farming.

4. Using your calculations from question 3, carry out a test of the hypothesis that the up-slope angles are 'no greater than' the down-slope angles.

5. In what ways could confidence in the above conclusions be improved?

6. Briefly outline one other hypothesis that could be tested with this data, and state the appropriate Student's t-test(s).

7. Briefly suggest some effective influences on the altitude of the moorland edge, which were not investigated in this study. The ways in which these influences might operate, and the methods that could be used to investigate them, should be mentioned.

10

χ^2 Tests and the Analysis of Contingency Tables

IN CHAPTER 9 Student's t-tests were used to test the significance of differences between two sample means, assuming interval scales of measurement and normal distributions. In this chapter somewhat less powerful but more widely applicable tests, based on the χ^2 (Chi-square) statistic, are introduced. Greater applicability is gained from the non-parametric nature of the tests—they do not require normal distributions—their ability to test differences between more than two samples simultaneously, and their less stringent requirements concerning the level of measurement of the data (ordinal-scale or nominal-scale data can be accommodated, as well as interval-scale data). The power of a χ^2 test, the ability of the test to detect significant differences, is somewhat lower than the power of the corresponding parametric test for the same reasons that make it of wider applicability. The limitations of the tests will be dealt with more fully at the end of the chapter.

One-sample tests

Although a χ^2 test can be applied to nominal, ordinal or interval-scale data, use of the test requires that the data be arranged in the form of a *contingency table*, which shows the number or frequency of occurrence of individual measurements or observations within categories or classes. An example of a contingency table is set out below, using data from the field of Medical Geography:

	Season				
	Spring	Summer	Autumn	Winter	Total
Observed frequency (O) (No. of cases of D.H.F.)	65	255	860	140	1320

The table shows the number of cases of dengue haemorrhagic fever (D.H.F.) in the four seasons of the year, for Singapore in 1973 (from Aiken and Leigh, 1978). This is a 1×4 contingency table with one sample and four categories. It should be noted that the occurrences or *observed frequencies* have been placed into categories on a nominal-scale level of measurement. If interval-scale or ordinal-scale data are to be used, then they too must be grouped into categories for purposes of this test.

The data suggest that dengue fever is most prevalent in the autumn season (August, September and October). A χ^2 one-sample test can be used to investigate the seasonality of the fever, and in particular to test the significance of the difference of the observed frequencies from the frequencies that would be expected if there were 'no difference' in the incidence of the fever through the seasons. The *expected frequencies* for no seasonality in the incidence of fever are:

	Spring	Summer	Autumn	Winter	Total
Expected frequency (E)	330	330	330	330	1320

The χ^2 test is essentially a method for determining whether the difference between the observed and expected frequencies are greater than are likely to have occurred by chance. Whether or not this is the case is determined by comparing a measure of the discrepancy between observed and expected frequencies (the calculated χ^2 statistic) with the discrepancy that is likely to occur by chance, as a result of sampling, at a given probability level (the tabulated χ^2 statistics, Table D, Appendix).

The calculated χ^2 statistic is given by the formula:

$$\chi^2 = \sum \frac{(O - E)^2}{E}$$

where O = an observed frequency,
      E = an expected frequency,
$(O - E)$ = the difference between an observed and an expected frequency.

For the example, four steps are involved in the calculation:

		Spring	Summer	Autumn	Winter
(a)	$(O - E)$	265	75	530	190
(b)	$(O - E)^2$	70,225	5625	280,900	36,100
(c)	$\dfrac{(O - E)^2}{E}$	212.80	17.05	851.21	109.39
(d)	$\sum \dfrac{(O - E)^2}{E}$	1190.45 = the calculated χ^2 statistic			

The appropriate tabulated χ^2 statistic is obtained from Table D (Appendix) using $(h - 1) = 3$ degrees of freedom, where h is the number of categories in the contingency table. Using a 5% significance level, the tabulated χ^2 statistic is 7.82. If a χ^2 value as large as 7.82 can be attributed to chance at the 5% significance level, then the calculated value of 1190.45 reflects a far greater difference than would be likely to occur if there was in fact 'no difference' between the observed data and a uniform distribution of dengue fever throughout the year. The hypothesis of 'no difference' must therefore be rejected. Use of the 5% significance level means that there is a less than 5% chance of having made a wrong decision; that is, we are greater than 95% certain of a real difference. Furthermore, it can be seen from Table D (Appendix) that the hypothesis of 'no difference' would be rejected at all the significance levels tabulated, so that we can be extremely confident in declaring there to be a seasonal pattern in the incidence of D.H.F. Although the example has involved testing the hypothesis of 'no difference' from a uniform distribution, it would be quite possible to test a hypothesis of 'no difference' from some other distribution of interest.

The χ^2 test may be viewed graphically with reference to Fig. 46(C). Possible calculated χ^2 values occur somewhere beneath the curve (a sampling distribution). If the calculated χ^2 exceeds the critical value, which is tabulated for particular significance levels, then the hypothesis of 'no difference' is rejected. The larger the calculated χ^2, the more likely that it will lie within the critical region for rejection of the 'null hypothesis'. The tabulated χ^2 statistic increases from zero towards the right in the figure and reflects the fact that the calculated χ^2 statistic increases as the differences between observed and expected frequencies increase (irrespective of the sign of the differences). χ^2 tests should therefore be viewed as one-tailed tests, but within the upper tail both positive and negative differences are represented. Occasionally, the χ^2 test is used to test for significant similarity; such tests involve the lower tail in Fig. 46(C) and the left-hand side of the χ^2 tables. In these tests the calculated χ^2 must be less than the tabulated χ^2 in order to *accept* the hypothesis of 'no difference'; one would use the 95% significance level in Table D (Appendix) if one wanted to be 95% sure of the hypothesis of 'no difference' being true.

Tests involving two or more samples and two or more categories

The last example was concerned with observed frequencies in four categories of one sample. Contingency tables involving more than one sample (k samples) and more than one category (h categories) can be analysed in a similar manner, providing a very flexible test, which is widely used in Geography and in other subjects for assessing the significance of differences between two or more samples. Here a worked example will be based on the aspect (orientation) of glaciers in the U.S.A. and the U.S.S.R. Figure 48 shows the orientation of a sample of glaciers in the North Cascades, Washington, and of a second sample from the Tashkent area, U.S.S.R. The data are represented as vector diagrams, which depict the number of glaciers orientated towards each of the eight points of the compass and indicate the resultant tendency for the glaciers to possess a north-north-east aspect (Evans, 1977). Such a preferred orientation might be explained in terms of a combination of prevailing snow-bearing winds and a shading effect (controls on the accumulation and melting of glacier ice). The example involves a χ^2 test of the hypothesis of 'no difference' between the two regions with respect to glacier orientation.

The data in Fig. 48 are first arranged in a 2×8 contingency table:

Observed frequencies (O)	Aspect category								
	SW	W	NW	N	NE	E	SE	S	Row total
N. Cascades (U.S.A.)	31	37	86	189	150	89	52	32	666
Tashkent area (U.S.S.R.)	6	9	69	103	109	40	35	3	374
Column total	37	46	155	292	259	129	87	35	1040

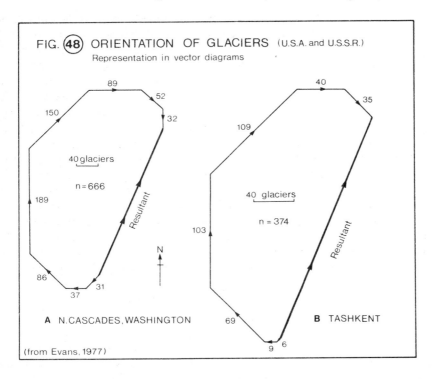

FIG. (48) ORIENTATION OF GLACIERS (U.S.A. and U.S.S.R.)
Representation in vector diagrams

A N.CASCADES, WASHINGTON

B TASHKENT

(from Evans, 1977)

Expected frequencies are next calculated, for each of the sixteen cells of the contingency table. Each expected frequency is obtained from the 'formula':

$$E = \frac{\text{Row total} \times \text{Column total}}{\text{Overall total}}.$$

Thus, for the N. Cascades and a south-westerly aspect the expected number of glaciers is $(666 \times 37)/1040 = 24$. For the other cells the expected frequencies are:

Expected frequency (E)	SW	W	NW	N	NE	E	SE	S
N. Cascades	24	29	99	187	166	83	56	22
Tashkent area	13	17	56	105	93	46	31	13

The expected frequencies are the frequencies that would be expected if there were indeed 'no difference' between the two regions in terms of glacier orientations; each E value taking into account the proportion of glaciers in each region and in each aspect category. Again, if 37 glaciers out of a total of 1040 glaciers

$$\left(\text{Proportion} = \frac{37}{1040} \right)$$

are known to face south-west, then in a region with 666 glaciers,

$$\text{the number expected is } 666 \times \frac{37}{1040}$$

(which is in agreement with the expected frequency derived from the 'formula').

The calculated χ^2 statistic is obtained using the same four steps as in the one-sample test, namely as shown in Table 10.

TABLE 10

		SW	W	NW	N	NE	E	SE	S	
(a)	$(O - E)$	7	8	13	2	16	6	4	10	N. Cascades
		7	8	13	2	16	6	4	10	Tashkent area
(b)	$(O - E)^2$	49	64	169	4	256	36	16	100	N. Cascades
		49	64	169	4	256	36	16	100	Tashkent area
(c)	$\dfrac{(O - E)^2}{E}$	2.04	2.21	1.71	0.02	1.54	0.43	0.29	4.55	N. Cascades
		3.77	3.76	3.02	0.04	2.75	0.78	0.52	7.69	Tashkent area

(d) $\sum \dfrac{(O - E)^2}{E} = 35.12 =$ calculated χ^2 statistic

The tabulated χ^2 statistic is obtained from Table D using $(k - 1)(h - 1) = 7 \times 1 = 7$ degrees of freedom ($h = 8$ categories, $k = 2$ samples). With a 5% significance level, the tabulated χ^2 statistic is 14.07. The calculated statistic therefore exceeds the value that would be likely to result by chance if the two samples of glaciers were in fact drawn from the same population. In other words, the hypothesis of 'no difference' between the two areas is rejected at the 5% significance level; there is a less than 5% chance (0.05 probability) of 'no difference' between the two regions, and we are greater than 95% certain of a difference (despite the apparent similarity between the two vector diagrams in Fig. 48). Use of a higher significance level (Table D) only confirms and emphasizes the confidence that can be placed in this decision.

Limitations of χ^2 tests

Although these tests do not possess the restrictive assumptions of the parametric tests, which accounts in part for their wide use in the context of geographical problems, they are not without limitations, and should not be applied indiscriminately. In particular, the following requirements should be met:

1. Data must be in the form of frequencies (that is, counted data within categories). χ^2 tests are eminently suited to comparing frequencies within nominal-scale categories, but can also be applied to higher-order levels of measurement if the data are grouped into categories prior to analysis. It is in this sense that these tests are not applicable to interval scale data.

2. The contingency table, in which the observed frequencies are placed, must consist of at least two categories (columns).

3. Expected frequencies in any cell of a contingency table should not be less than 5. Although it is permissible for 20% of cells to have expected frequencies of less than 5 when contingency tables are larger than 2×2, no cell is allowed to have an expected frequency of less than 1. This requirement can sometimes be met by the amalgamation of categories, producing fewer cells with more observations in each.

4. Samples are assumed to be independent. The test is therefore not applicable to dependent samples.

5. Random sampling is assumed. This assumption is common to all the statistical tests contained in the manual but it is considered valid to use other sampling designs provided that they are unbiased (see also Chapter 15).

Exercise 14: Hypothesis testing about the location of Eskimo settlements on the coast of Baffin Island, Canada, using χ^2 tests.

Background

Archaeological and historical evidence has revealed that the eastern coast of Baffin Island was occupied by Eskimo populations over much of the last 2000 years. As well as temporary summer tent rings, there are abundant remains of more permanent winter dwellings. The location of forty-two summer sites and fifty-five winter sites is shown in Fig. 49. The winter sites have been assigned, on the basis of type of structure, to three cultural stages: first, the 'Early Thule' culture (A.D. 1200 to 1550), characterized by semi-subterranean turf and stone houses; second, the 'Late Thule' culture (A.D. 1550 to 1850), indicated by turf and stone qarmats (tent foundations); and third, the 'Historic' culture (post-A.D. 1850) with wood and canvas frame houses. Although some of the summer sites can be assigned to the 'Historic' cultural stage, most of these are of uncertain age.

The climate of the region is very severe with a short summer season. Mean summer temperatures are about $-10°C$, with average daily temperatures rising above zero from mid-June to the end of August in most years. Coastal ice-floes add to the difficulties of life and in some summers may not break up at all. Biological productivity is low on land and the Eskimo cultures were geared to the maritime ecosystem (particularly to the ringed seal). Despite the climatic fluctuations of the last 2000 years, including the 'Little Ice Age', very severe climatic conditions have characterized the period of interest.

It is the purpose of the present exercise to test hypotheses relating to the Eskimo dwellings, and to characteristics of their sites, using χ^2 tests. In this way it should be possible to make an overall assessment of whether or not the Eskimo populations were adapted to their environment in this region of climatic stress. The data are given in Table 11, which, in addition to cultural stage and seasonal character of the dwellings, indicates the following site characteristics: (a) aspect, according to four points of the compass; (b) presence or absence of a protected beach; (c) presence or absence of land rising to greater than 150 m behind the dwelling; and (d) degree of development of the soil/vegetation complex, according to three categories.

Practical work

1. What is the level of measurement of the following characteristics of the sites of Eskimo dwellings:
   (a) Aspect?
   (b) Presence or absence of land above 150 m behind the dwellings?
   (c) Presence or absence of a protected beach?
   (d) Degree of development of the vegetation/soil complex?

TABLE 11. *Site characteristics of Eskimo settlements Baffin Island, Canada*

Site no.	Stage	Season	Aspect	Height	Beach	Soil
1	?	S	W	+	–	1
2	H	S	S	+	+	1
3	?	S	S	+	+	1
4	H	W	S	+	+	1
5	ET	W	S	+	+	2
6a	ET	S	S	+	–	1
6b	H	W	S	+	–	2
7a	ET	S	S	+	+	2
7b	H	W	W	+	+	2
8a	ET	S	W	+	+	3
8b	?	W	S	+	+	3
9	LT	S	S	+	+	1
10	?	S	W	+	+	1
11	?	S	N	–	–	1
12	?	W	E	–	–	1
13a	LT	S	E	–	–	1
13b	?	S	S	–	+	1
14	H	W	S	–	–	2
15	?	S	S	–	+	2
16a	LT	W	S	+	+	2
16b	?	S	S	+	+	2
17	LT	W	S	+	+	2
18a	ET	S	S	+	+	2
18b	?	W	S	+	+	2
19	H	S	N	–	+	2
20a	LT	W	N	–	–	1
20b	?	S	N	–	–	1
21a	ET	W	S	–	–	2
21b	?	S	S	–	–	2
22a	ET	W	S	+	–	2

Site no.	Stage	Season	Aspect	Height	Beach	Soil
36b	?	S	W	+	–	2
37	LT	W	W	–	+	1
38a	LT	W	S	+	+	1
38b	?	S	S	+	+	1
39a	LT	W	S	+	+	1
39b	?	S	S	+	+	1
40	H	W	S	+	+	1
41a	ET	W	S	+	+	2
41b	LT	S	S	+	+	2
41c	?	W	S	+	+	2
42a	ET	W	S	+	–	2
42b	LT	W	S	+	–	3
42c	?	S	S	+	–	3
43a	ET	W	E	+	+	3
43b	H	W	E	+	+	3
43c	?	S	E	+	+	3
44a	ET	W	E	+	+	1
44b	LT	W	E	+	+	1
44c	?	S	E	–	+	1
45a	ET	W	E	–	+	2
45b	H	S	E	–	+	2
45c	?	W	E	+	+	2
46	?	W	E	+	+	1
47	ET	S	S	+	+	2
48a	H	W	E	+	+	1
48b	?	W	E	+	+	1
49	H	S	S	+	+	3
50	LT	W	S	+	–	2
51	LT	W	S	+	+	1
52a	ET	W	S	+	+	2

Site	Stage	Season	Aspect	Height	Beach	Soil
22b	?	S	S	+	−	2
23a	ET	W	S	+	+	3
23b	?	S	S	+	+	3
24	LT	W	S	+	+	1
25	ET	W	S	+	+	3
26	ET	W	S	+	+	2
27	ET	W	W	+	+	2
28	ET	W	S	+	+	2
29a	ET	W	S	+	+	2
29b	?	W	W	+	+	3
30	ET	S	W	+	+	1
31	?	W	W	+	+	1
32a	ET	W	W	+	+	1
32b	LT	W	W	+	+	2
33	ET	W	S	+	+	3
34a	LT	S	S	−	−	3
34b	?	S	S	−	−	1
35	LT	W	S	+	+	2
36a	LT	W	W	+	−	2
52b	LT	W	S	+	+	2
52c	?	S	S	+	+	3
53a	LT	W	N	+	+	3
53b	H	S	N	+	+	1
54	?	S	S	+	−	3
55	?	S	S	+	+	2
56	?	S	E	+	+	2
57a	ET	W	S	+	+	2
57b	LT	W	S	+	+	2
57c	H	W	S	+	+	3
58	LT	S	S	+	+	1
59a	LT	W	S	+	+	1
59b	H	W	E	+	+	1
60	H	W	S	+	+	2
61	H	W	W	+	+	3
62	LT	S	W	−	−	3
63a	ET	W	W	+	+	1
63b	?	S	W	+	+	2

Key: Stage = Cultural Stage (ET = Early Thule, LT = Late Thule, H = Historic, ? = Unknown).
Season = Seasonal group (S = summer dwelling, W = winter dwelling).
Aspect = Aspect of site (N = north, S = south, E = east, W = west).
Height = Presence (+) or Absence (−) of land over 150 m behind the dwelling.
Beach = Presence (+) or Absence (−) of a protected beach.
Soil = Degree of development of the vegetation/soil complex (1 = thin or absent soil and discontinuous or absent vegetation, 2 = developed soil and discontinuous or absent soil and continuous vegetation, 3 = deep soil and lust vegetation).

(From Jacobs and Sabo, 1978.)

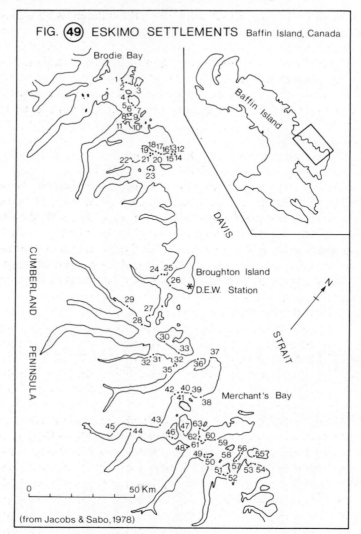

FIG. ㊾ ESKIMO SETTLEMENTS Baffin Island, Canada

(from Jacobs & Sabo, 1978)

2. Suggest ways in which each of the above characteristics of the environment would be anticipated to influence the selection of dwelling places by Eskimos.

3. Using the data for all dwellings, irrespective of cultural stage or seasonal character, use χ^2 tests to help answer the following questions:

   (a) Do the dwelling places have a preferred aspect?

   (b) Are the dwelling places associated with protected beaches?

   (c) Are the dwelling places related to the degree of development of the vegetation/soil complex?

Each answer should include a full account of the χ^2 test, including a contingency table, an explicit statement of the hypothesis of 'no difference' (null hypothesis) that is actually tested, a clear explanation of the decision reached, and comments on any limitations of the test.

4. Use χ^2 tests to investigate the significance of the differences between winter and summer sites in respect of the following:

(a) Aspect of the site.

(b) The presence or absence of land over 150 m behind the dwelling.

(c) The quality of vegetation and soil at the sites.

5. Basing your answers on χ^2 tests, are there grounds for suggesting that dwelling places of the 'Early Thule' culture differ from those of the 'Late Thule' culture in terms of being associated with:

(a) Protected beaches?

(b) Land rising to over 150 m behind the dwellings?

(c) Degree of development of vegetation/soil at the site?

6. Write an essay on the degree to which the Eskimo culture can be said to have been adapted to its environment. Your answer should be based on the χ^2 tests that you have carried out and should pay particular attention to any differences detected between cultural stages and between seasonal categories of dwelling place.

7. Discuss the relative merits of an approach based on χ^2 tests and an approach based on a qualitative appraisal of the data. (It should be noted that the research paper by Jacobs and Sabo (1978), on which this exercise is based, merely described the tabulated data, without the application of statistical tests.)

Exercise 15: Analysis of spatial variation of water quality in the River Exe drainage basin (south-west England) using χ^2 tests.

Background

Water quality constitutes a valuable environmental indicator, the dissolved load of a river reflecting, amongst other factors, the inputs to the basin from precipitation, the geological strata over which the river is flowing, and the land-use practices in operation on the surrounding slopes. In this exercise, the effect of land use on the water quality of tributaries of the River Exe is examined, while controlling for the effect of geology. In this way it is possible to assess whether land use is an effective input to the spatial variation in dissolved load.

The basin of the River Exe extends over a variety of rock types, the three most widespread being shown in Fig. 50(A). Water samples were taken at over 500 sites in early June 1971 during several days of dry weather with stable low river levels. Accessible points were chosen, usually on small tributaries, so as to ensure a relatively rapid and even spatial coverage of the entire basin (Fig. 50B). Using a conductivity meter, the specific conductance of each sample was measured in micromhos/cm, thus providing an accurate set of near-contemporaneous measurements of total dissolved load over the whole basin. The predominant land use (woodland, farmland, or moorland) was noted for the catchment area of each sampling point.

Initial inspection of the data revealed that rock type was the primary control on river water quality in the Exe basin. This is demonstrated in Fig. 51, which shows histograms of the specific conductance values grouped according to rock type. These groups were compared using χ^2 tests and were found to be statistically different at the 1.0% significance level (Walling and Webb, 1975). The susceptibility to weathering of the various rock types

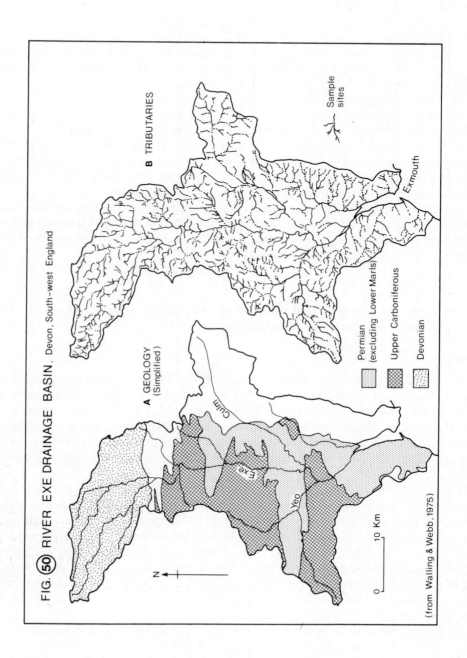

FIG. ㊿ RIVER EXE DRAINAGE BASIN, Devon, South-west England

A GEOLOGY (Simplified)

B TRIBUTARIES

Permian (excluding Lower Marls)

Upper Carboniferous

Devonian

Sample sites

Exmouth

Culm

Exe

Yeo

N

0   10 Km

(from Walling & Webb, 1975)

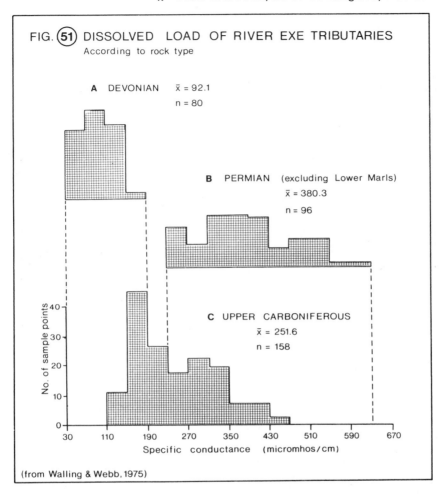

FIG. (51) DISSOLVED LOAD OF RIVER EXE TRIBUTARIES
According to rock type

A   DEVONIAN $\bar{x} = 92.1$
                 $n = 80$

B   PERMIAN   (excluding Lower Marls)
              $\bar{x} = 380.3$
              $n = 96$

C   UPPER CARBONIFEROUS
    $\bar{x} = 251.6$
    $n = 158$

No. of sample points

Specific conductance   (micromhos/cm)

(from Walling & Webb, 1975)

controls the input of dissolved load to the various tributaries. The Middle and Upper Devonian strata are siliceous and weather only slowly, resulting in a mean specific conductance of 92.1 micromhos/cm; at the other extreme, the Permian rocks weather more rapidly and deeply, and are characterized by much higher conductances (mean = 380.3 μmhos/cm); the Upper Carboniferous rocks are intermediate in character (mean conductance = 251.6 μmhos/cm). If a land-use effect is to be detected, therefore, the influence of geology must be in some way controlled. The approach adopted here is to use the data from the three rock types as stratified samples. In other words, data from each rock type are analysed separately (by use of χ^2 tests) for the effect of land use.

Practical work

The following tables (Tables 12a-c) show the frequency of occurrence of particular levels of specific conductance (measured in μmhos/cm) found in tributaries associated with various combinations of rock type and land use.

TABLE 12a. *Middle and Upper Devonian tributaries*

	Specific conductance class			
	30–69	70–109	110–149	150–189
Woodland	3	9	12	0
Farmland	0	13	12	2
Moorland	20	8	1	0

(From Walling and Webb, 1975)

TABLE 12b. *Permian tributaries (excluding lower Marls)*

	230–269	270–309	310–349	350–389	390–429	430–469	470–509	510–549	550–589	590–629
Woodland	4	1	3	4	3	1	2	0	0	0
Farmland	9	7	13	13	13	5	8	8	1	1

(From Walling and Webb, 1975)

TABLE 12c. *Upper carboniferous tributaries*

	110–149	150–189	190–229	230–269	270–309	310–349	350–389	390–429	430–469
Woodland	10	33	9	3	7	7	3	3	2
Farmland	1	12	17	14	17	12	4	4	0

(From Walling and Webb, 1975)

1. Throughout this and subsequent questions care should be taken to ensure that the assumptions of χ^2 tests are met, before proceeding with the calculations. It may be found that Tables 12a–c cannot be used in their present form, but can be used if some regrouping is carried out.

   (a) Using the data for the Middle and Upper Devonian rock type, and a single χ^2 test, is there a significant difference between the dissolved load of the tributaries associated with different land uses?

   (b) Test the significance of the differences between each pair of land uses, using the data relating to Middle and Upper Devonian rocks.

   (c) Discuss the added information gained from carrying out the tests in 1(b).

For each test, the hypothesis of 'no difference' should be stated, a significance level selected, and the statistical decision clearly explained.

2. As far as you are able, use the data relating to the other two rock types to test further the conclusions reached in question 1.

3. To what extent do the results of the χ^2 tests support the following statements:

   (a) Land use differences cause differences in the dissolved load of rivers?

   (b) The effect of farming on the quantity of dissolved load in rivers is greater than the effect of forestry?

   (c) If an area of moorland is subjected to afforestation, this will result in an increase in the dissolved load of the catchment?

4. Briefly discuss the ways in which land-use practices may indeed alter the dissolved load of rivers.

5. In the background information to the present exercise it was pointed out that the researchers who carried out this study found statistically different dissolved loads to be associated with different rock types (Fig. 51 illustrates these differences in visual form). Bearing in mind the data shown in Fig. 51, and your experiences so far with χ^2 tests, critically assess how this may have been done.

11

Further Non-parametric Tests for Independent Samples

THIS chapter considers three non-parametric statistical tests. These are suitable for testing differences between two or more independent samples, and all three tests require at least an ordinal-scale level of measurement. They have, therefore, more exacting data requirements than the χ^2 tests, but less exacting requirements than the parametric Student's t-tests. When the assumptions of χ^2 tests or of Student's t-tests are not met, one of the following may be applicable.

The Kolmogorov–Smirnov two-sample test

The Kolmogorov–Smirnov two-sample test is a test of the difference between two independent samples. It is therefore useful in the same sort of situation as a two-sample χ^2 test, although it avoids what is perhaps the main limitation of χ^2 tests—namely the requirement of sufficient observations within each cell of the contingency table to give rise to sufficiently high expected frequencies. The Kolmogorov–Smirnov test circumvents this difficulty by comparing two cumulative frequency histograms, rather than two histograms (the latter being a way of visualizing the equivalent χ^2 test). If the two cumulative frequency histograms differ by more than is likely to occur by chance as a result of sampling from the same population, then the hypothesis of 'no difference' can be rejected.

The test requires a very simple calculation; the calculation of the D statistic, which is the *maximum difference between the two cumulative frequency histograms* (when both histograms are expressed as proportions or percentages). The tabulated D statistic, describing the highest value of D that can be expected to occur by chance at a known probability level, is given in Table E (Appendix). A worked example from a study of perception and attitudes towards water management in south-west Ontario, Canada, is detailed below.

A number of questions were directed towards the 'public' and towards 'professionals' to find out whether these two groups (samples) differed in their perception of a range of water-management problems (Mitchell, 1971). One of the questions asked of a sample of 400 members of the public and a sample of 40 professionals was:

"The average person may not know what is best for him where technical problems are concerned and should rely upon professionals."

Do you (a) Strongly Agree?
(b) Agree?
(c) Neutral?
(d) Disagree?
(e) Strongly Disagree?

The results are summarized in the following table, in which the percentage of the public and the percentage of the professionals who answered this question are indicated:

	Strongly Agree	Agree	Neutral	Disagree	Strongly Disagree
Public	9	80	7	4	0
Professionals	18	55	8	16	3

The application of the Kolmogorov–Smirnov test involves a test of the hypothesis of 'no difference' between the responses of the two groups. The results are therefore recast in the form of cumulative percentage frequencies:

	Strongly Agree	Agree	Neutral	Disagree	Strongly Disagree
Public	9	89	96	100	100
Professionals	18	73	81	97	100

The calculated D statistic is thus 0.16, for 16 % is the maximum difference between the two sets of cumulative percentage frequencies. This is shown in the form of a histogram in Fig. 52. The tabulated D statistic is obtained from Table E, and using the 5 % significance level:

$$D = 1.36 \sqrt{\frac{n_1 + n_2}{n_1 n_2}}$$

where n_1 = sample size of sample 1 (public) = 400 and n_2 = sample size of sample 2 (professionals) = 40. The tabulated D statistic is thus

$$D = 1.36 \sqrt{\frac{440}{16,000}} = 0.2255 \text{ or } 22.55\%.$$

The hypothesis of 'no difference' cannot in this case be rejected, because the calculated difference does not exceed the difference that is likely to occur by chance. In other words, we are less than 95 % confident in the existence of a real difference between the attitude of the public and the attitude of the professionals over this question; that is, there is a greater than 5 % chance of 'no difference'.

The following limitations to the Kolmogorov–Smirnov two-sample test should be noted:

1. Unless sample sizes are equal, Table E can only be used when both sample sizes are greater than about 40.

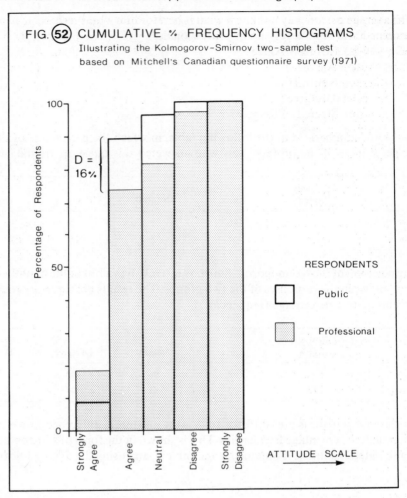

FIG. ⑤② CUMULATIVE % FREQUENCY HISTOGRAMS
Illustrating the Kolmogorov–Smirnov two–sample test
based on Mitchell's Canadian questionnaire survey (1971)

2. At least two categories (and preferably as many as the data permit) must be available
3. The categories must be at least ordinal-scale categories.
4. The samples are assumed to be independent.
5. Random sampling is assumed.

The Mann–Whitney two-sample test

This test is used for the same purpose as the Kolmogorov–Smirnov test, that is to test whether two independent samples could have been drawn from the same population. Whereas both tests (and the two-sample χ^2 test) are sensitive to any kind of difference between the samples (such as central tendency, dispersion or skewness) the Mann – Whitney test is most sensitive to differences in central tendency. It can therefore be regarded as the most powerful alternative to the Student's t-test of a difference between two sample means, and is most appropriate when the assumptions of the Student's t-test are not met. Like the Kolmogorov–Smirnov test, the Mann–Whitney test is applicable to very small samples as well as to large ones; the former test is slightly more powerful with small samples, but the latter can be applied to samples with unequal sample sizes.

In a Mann–Whitney test the individual observations or measurements are first listed in rank-order, irrespective of whichever of the two samples they belong to. If there was, in reality, 'no difference' between the populations from which the samples were drawn, then the sum of the ranks of the observations in sample 1 would be approximately equal to the sum of the ranks in sample 2. The hypothesis of 'no difference' can be rejected if the sum of the ranks of either sample differs sufficiently from the sum of the ranks that is likely to occur by chance. The test involves the calculation of the Mann–Whitney U statistic, which is dependent on the sum of the ranks of one of the samples. The tabulated U statistic, describing the highest value of U that can be expected to occur by chance at a known probability level, is given in Table F (Appendix).

As a worked example, the data from Exercise 12 will be used, and the number of species present on ground in the first two time-zones will be compared. The data, which represent the number of species per 16 m^2, are set out in Table 13, together with the rank-order of each measurement. It should be noted that the data are the same as those in Exercise 12, but they have been rearranged in order of increasing value (within each sample) and have been ranked irrespective of the sample in which they occur. The lowest value has been given rank 1.

TABLE 13

Sample 1 ($n_1 = 16$)		Sample 2 ($n_2 = 30$)	
Value	Rank	Value	Rank
2	1	4	3
4	3	7	15.5
4	3	7	15.5
5	5	8	19
6	9.5	8	19
6	9.5	10	23.5
6	9.5	10	23.5
6	9.5	10	23.5
6	9.5	10	23.5
6	9.5	11	27
6	9.5	11	27
6	9.5	11	27
7	15.5	12	30
7	15.5	12	30
8	19	12	30
9	21	13	33
		13	33
		13	33
		14	35.5
		14	35.5
		15	37.5
		15	37.5
		16	39.5
		16	39.5
		17	42
		17	42
		17	42
		18	44
		19	45
		24	46

The hypothesis of 'no difference' (null hypothesis) to be tested is that there is 'no difference' between the two samples at the 5 % significance level. The sum of the ranks is calculated for each sample and found to be $R_1 = 159.0$ and $R_2 = 922.0$. Note that some of the ranks are ties; when a tie occurs, the average rank is used. The calculated U statistic is whichever of the following two values is the greater:

$$\text{(a)} \quad U' = n_1 n_2 + \frac{n_1(n_1 + 1)}{2} - R_1$$

$$= 16.30 + \frac{16(16 + 1)}{2} - 159$$

$$= 480 + 136 - 159 = 457$$

$$\text{or} \quad \text{(b)} \quad U'' = n_1 n_2 - U' = 480 - 457 = 23$$

where n_1 = the sample size of sample 1,
$\quad\quad\quad n_2$ = the sample size of sample 2,
$\quad\quad\quad R_1$ = the sum of the ranks of sample 1.

In this case, therefore, the calculated U statistic is 457.

Referring to Table F (Appendix) we find that a value of U as large as 326 would be expected to occur by chance at the 5 % significance level. As the calculated U statistic is 457, the difference between the two samples is greater than can be attributed to chance, so that the hypothesis of 'no difference' must be rejected. In other words there is a greater than 95 % chance of a real difference between the two time zones in terms of number of species. Indeed, the hypothesis of 'no difference' is also rejected at the 1 % significance level, which means that there is a less than 1 % chance of 'no difference'.

When sample sizes are greater than those shown in Table F (Appendix) the significance of U may be found by use of tables of the normal distribution function, z being found from U by means of the formula:

$$z = \frac{U - \dfrac{n_1 n_2}{2}}{\sqrt{\dfrac{(n_1)(n_2)(n_1 + n_2 + 1)}{12}}}$$

In the present example, z becomes:

$$\frac{457 - \dfrac{480}{2}}{\sqrt{\dfrac{480(16 + 30 + 1)}{12}}} = \frac{217}{\sqrt{\dfrac{(480)(47)}{12}}} = \frac{217}{43.359} = 5.0$$

It should be noted that tables of the normal distribution function (Table A, Appendix) are arranged in a form suitable for one-tailed tests. In order to reject a hypothesis of 'no difference' at the 5 % significance level a tabulated z of 1.96 must be exceeded. The Mann–Whitney test has the following limitations:

1. At least ordinal scale data are required.
2. If the number of ties between samples is very large, then the test may be affected. The effect of ties is usually negligible, however, and tied ranks within a sample do not influence the results.
3. The samples are assumed to be independent.
4. Random sampling is assumed.

The Kruskal–Wallis test for more than two samples

Three or more (k) independent samples may be compared using the Kruskal–Wallis one-way analysis of variance by ranks. This test involves a hypothesis of 'no difference' between more than two samples and tests whether the k samples are likely to have been drawn from the same population. It is applicable in situations where neither of the other two tests discussed in this chapter can be used. Moreover, it is more powerful than the equivalent χ^2 test as it utilizes more information, being based on ordinal-scale data.

The Kruskal–Wallis test is related in principle to the Mann–Whitney test in that it requires the calculation of a statistic based on the sum of the ranks for each sample, when all observations (irrespective of the sample to which they belong) have been placed in rank-order. If all the k samples were drawn from the same population, then it would be expected that the sum of the ranks for each sample would be approximately the same. The calculated Kruskal–Wallis H statistic is a measure of the actual difference in the sum of the rankings between samples. If this measured difference exceeds the difference that is likely to have occurred by chance, then the hypothesis of 'no difference' between the k samples can be rejected.

The calculated H statistic is obtained from the formula:

$$H = \frac{12}{N(N+1)} \left[\sum \frac{R^2}{n} \right] - 3(N+1)$$

where N = the overall, total sample size (that is, the sum of the sample sizes of all k samples),
  n = the sample size of a particular sample,
  R = the sum of the ranks of a particular sample,
  R^2 = the square of R,
  $\sum \frac{R^2}{n}$ = for each of the k samples, the quantity R^2/n is calculated, and the sum of these k quantities is taken.

When there are three samples or less, and when sample sizes are 5 or less ($n \leqslant 5$), then Table G (Appendix) gives the tabulated H statistic. For a greater number of samples and sample sizes, the tabulated χ^2 statistic can be used with $(k-1)$ degrees of freedom (Table D, Appendix).

The worked example tests the difference in shape of particles in four types of deposit in the Oetztal Alps of Austria. The data tabulated in Table 14 are values of a 'flatness' index for particles from the four types of deposit: (i) glacial till; (ii) fluvio-glacial sediments; (iii) supra-glacial debris; and (iv) scree. The flatness index was calculated from the dimensions of the particles, as indicated in Fig. 53A.

TABLE 14

Till		Fluvio-glacial		Supra-glacial		Scree	
Value	Rank	Value	Rank	Value	Rank	Value	Rank
1.1	1	1.5	11	1.5	11	1.4	5.5
1.2	2	1.5	11	1.8	25	1.6	16.5
1.3	3	1.5	11	2.0	35	2.0	35
1.4	5.5	1.6	16.5	2.0	35	2.0	35
1.4	5.5	1.6	16.5	2.1	41	2.5	52
1.4	5.5	1.7	21	2.2	45	2.6	57
1.5	11	1.7	21	2.5	52	2.6	57
1.5	11	1.8	25	2.7	60	2.8	61
1.5	11	1.8	25	3.0	65	3.0	65
1.6	16.5	1.9	29	3.0	65	3.1	67
1.7	21	1.9	29	3.2	68	3.5	74.5
1.7	21	1.9	29	3.3	70	3.8	77.5
1.7	21	2.1	41	3.4	72	4.1	82.5
1.9	29	2.1	41	3.5	74.5	4.1	82.5
1.9	29	2.1	41	3.5	74.5	4.1	82.5
2.0	35	2.2	45	3.8	77.5	4.4	86
2.0	35	2.3	48	3.9	79	5.4	87
2.0	35	2.3	48	4.0	80	5.7	88
2.1	41	2.5	52	4.1	82.5	6.5	90.5
2.2	45	2.6	57	4.3	85	6.5	90.5
2.3	48	2.6	57	6.1	89	7.4	94
2.5	52	2.9	62.5	6.7	92.5	8.4	95
2.5	52	2.9	62.5	6.7	92.5	9.1	97
2.6	57	3.3	70	8.5	96	10.2	99
3.3	70	3.5	74.5	9.2	98	10.8	100

(After Shakesby, pers. comm.)

The hypothesis to be tested is that there is 'no difference' between the four samples. The first step is to calculate the sum of the ranks for each sample:

$$R_1 = 663.0 \qquad R_2 = 944.5 \qquad R_3 = 1665.0 \qquad R_4 = 1777.5$$

R^2/n is calculated for each sample:

$$17,582.76 \qquad 35,683.21 \qquad 110,889.0 \qquad 126,380.24$$

$\sum R^2/n$ is thus the sum of these four quantities $= 290,535.21$
The calculated H statistic is therefore:

$$H = \frac{12}{100(100+1)} \times 290,535.21 - 3(100+1)$$

$$= (0.001\,188\,11 \times 290,535.21) - 303$$
$$= 345.184\,88 - 303$$
$$= 42.184\,88$$

Reference to Table D (Appendix) with $(k-1) = 3$ degrees of freedom yields a χ^2 value of 7.82 (using the 5 % significance level). As the calculated statistic exceeds the tabulated statistic, the measured difference can be said to be greater than would be expected if the four samples were drawn from the same population. The hypothesis of 'no difference' must

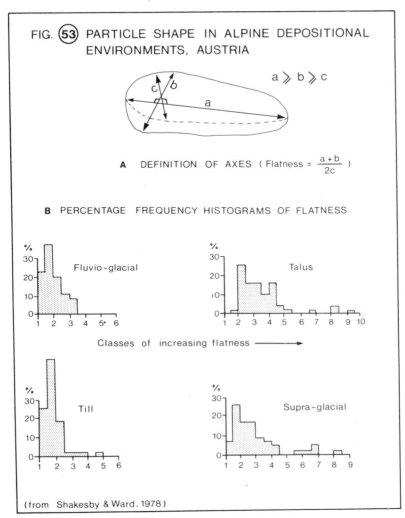

FIG. (53) PARTICLE SHAPE IN ALPINE DEPOSITIONAL ENVIRONMENTS, AUSTRIA

$a \gg b \gg c$

A DEFINITION OF AXES (Flatness = $\frac{a+b}{2c}$)

B PERCENTAGE FREQUENCY HISTOGRAMS OF FLATNESS

Fluvio-glacial

Talus

Classes of increasing flatness ⟶

Till

Supra-glacial

(from Shakesby & Ward, 1978)

therefore be rejected, and we can be at least 95 % sure of a real difference between the flatness of the particles comprising the four deposits.

The test indicates that there is a significant difference between the shape of the particles within the four depositional environments. No indication is given, however, as to the precise nature of the differences. The promising result from the Kruskal–Wallis test could be investigated further by more detailed analysis of differences between particular pairs of samples. This latter approach was adopted by Shakesby and Ward (1978) who used the Kolmogorov–Smirnov test to examine, for example, the possibility of one type of deposit being derived from another. Figure 53B shows the percentage frequency histograms for the larger data set on which the tests were carried out.

The limitations of the Kruskal–Wallis test are the same as those listed for the Mann–Whitney test.

Exercise 16: A comparative study of hobby farms and commercial farms in Ontario, Canada, using Kolmogorov–Smirnov tests.

Background

Hobby farms have been defined as 'a condition of land ownership in which a farm and residence have been added to non-farm employment' (Layton, 1978). Since the Second World War, hobby farmers have become a major land-owning group in many industrialized countries, but they are known from the rural estates of ancient China and Egypt. Some hobby farms are run essentially as recreational pursuits, whereas others are run primarily to supplement an income received elsewhere. In this exercise, an attempt is made to define some of the characteristics of hobby farms, using data collected from a postal questionnaire survey carried out in the rural–urban fringe of London, Ontario, Canada, an area of high quality agricultural land and rapid urbanization.

Hobby farming has increased rapidly since the 1960s in the thirteen townships involved in the survey, and at present hobby farms make up 13 % of the study area. Two other types of land ownership were recognized in the survey: first, full-time farms, with farming as the only occupation of the owner; and secondly, part-time farms, which have added non-farm employment to farm income. In addition, the hobby farms have been subdivided into two groups: first, the commercially-motivated group, who view hobby farming as a transitional step towards farming on a more-or-less commercial basis; and secondly, the non-motivated group, who regard hobby farming as recreation and attach little importance to the supplementary income obtained from farming.

The exercise uses the Kolmogorov–Smirnov two-sample test to examine the differences between the various groups in terms of farm size, their man-day requirement (intensity of operation), income received and the enterprises involved. In this way insights are obtained into the economic and social role of the hobby farm in the rural–urban fringe. It should be noted that the data below are in percentage form. Where more than two samples are available for comparison, the results are given of χ^2 tests carried out by Layton (1978) on the data in the form of frequencies.

Practical work

1. Using the data relating to farm size, use a Kolmogorov–Smirnov test to examine whether or not the differences between the following are significant:
   (a) hobby farms and part-time commercial farms;
   (b) motivated and non-motivated hobby farms;
   (c) part-time and full-time commercial farms.
Each test should be fully described at all stages.

2. With reference to the data on farm size, which of the following could be compared legitimately using a Kolmogorov–Smirnov test:
   (a) hobby farms and non-motivated hobby farms?;
   (b) motivated hobby farms and part-time commercial farms?;
   (c) hobby farms and commercial farms (full- and part-time combined)?
Do not attempt any calculations, but point out any aspects of the data that would prevent the valid application of a Kolmogorov–Smirnov test.

3. Referring to the data for farm income, and assuming the use of a Kolmogorov–Smirnov test, which of the possible comparisons between ownership types would be expected to yield:

TABLE 15. *Characteristics of hobby farms and commercial farms in Ontario, Canada*

A. *Percentage of farms of various sizes according to ownership type*

Ownership type	Farm size (acres)						
	< 10	10–25	26–50	51–100	101–200	> 200	(n)
Hobby	3	23	16	27	27	4	(114)
Part-time commercial	0	3	8	30	29	30	(109)
Full-time commercial	1	1	1	10	43	44	(187)
(a χ^2 test of a difference between the three ownership types was found significant at the 0.01 % level)							
Motivated hobby	0	6	9	35	38	12	(50)
Non-motivated hobby	5	35	16	18	24	2	(64)

(After Layton, 1978.)

B. *Percentage of farms with various man-day requirements according to ownership type*

Ownership type	Standard man-day requirement						
	0–24	25–49	50–99	100–199	200–365	> 365	(n)
Hobby	33	31	17	12	2	5	(114)
Part-time commercial	10	14	13	19	10	34	(109)
Full-time commercial	0	0	4	10	14	72	(187)
(a χ^2 test of a difference between the three ownership types was found significant at the 0.01 % level)							
Motivated hobby	10	35	20	20	15	0	(50)
Non-motivated hobby	61	24	9	3	0	3	(64)

(After Layton, 1978.)

C. *Percentage of farms with various incomes according to ownership type*

Ownership type	Gross income (dollars)					
	< 50	50–1999	2000–4999	5000–9999	> 10,000	(n)
Hobby	16	32	28	7	17	(114)
Part-time commercial	0	9	16	18	57	(109)
Full-time commercial	0	1	4	7	88	(187)
(a χ^2 test of a difference between the three ownership types was found significant at the 0.01 % level)						
Motivated hobby	3	16	31	20	30	(50)
Non-motivated hobby	40	32	7	11	10	(64)

(After Layton, 1978.)

D. *Percentage of farms in various enterprises according to ownership type*

Enterprise	Ownership type				
	Hobby	Part-time commercial	Full-time commercial	Motivated hobby	Non-motivated hobby
Non-commercial	29	1	0	52	6
General mixed	23	35	29	7	41
Cash crop	19	24	9	15	22
Beef	9	12	8	7	13
Horses	6	0	0	9	3
Mixed cash crops	5	7	6	7	3
Mixed livestock	5	6	10	0	9
Pigs	2	3	4	0	3
Horticulture	2	2	1	3	0
Poultry	0	1	3	0	0
Tobacco	0	1	5	0	0
Dairy	0	8	25	0	0

(After Layton, 1978.)

(a) the most statistically significant difference?;

(b) the least statistically significant difference?

Again, do not attempt the calculations, but give a brief reasoned argument for your choice in each case.

4. Using the data on man-day requirement, farm income and farm size, can one differentiate between the characteristics of the motivated hobby farm and the part-time commercial farm? This question requires the application of further Kolmogorov–Smirnov tests.

5. Discuss how the data on farm enterprises might be analysed to yield information on the differences between hobby farms and commercial farms. Your discussion should consider whether and how Kolmogorov–Smirnov and/or χ^2 tests could be applied, and what further information would be required (if any) in order to carry out the tests.

6. Write a short essay on the similarities and differences between hobby farms and commercial farms referring, where necessary, to your results and conclusions from questions 1–5.

Exercise 17: Do regions exist? Testing the distinctiveness of planning regions in South Wales using Kruskal–Wallis analysis of variance by ranks.

Background

Definitions of region(s) are numerous and diverse, as revealed by the following selection:

1. 'Any circumscribed territorial unit' (Rodoman).
2. 'An area whose physical conditions are homogeneous' (Joerg).
3. 'A geographic area unified culturally . . .' (Young).
4. 'An area wherein there has grown up one characteristic human pattern of adjustment to environment' (American Society of Planning Officials).
5. 'Genuine entities, each of which experiences both natural and cultural differentiation from its neighbours' (Reiner).
6. 'An area within which a higher degree of mutual dependency exists than in relationships outside that area' (Stanberry).
7. 'An area or unit within which the economic and social activity of the population are integrated around a focal and administrative centre' (Mackenzie).
8. 'A complex created by man and which man can destroy' (Gottmann).
9. 'A constellation of communities' (Dawson and Gettys).
10. 'A way of life' (Labasse).
11. 'Any one part of a national domain sufficiently unified . . . to have a true consciousness of its own customs and ideals, and to possess a sense of distinction from other parts . . .' (Royce).
12. 'Spatial structures which are smaller in area than the state, which possess a certain individuality, and which are considered as entities either by the people who live there or by outside observers' (Claval).

It is clear that different types of region are recognized and that there is no agreement on a universal definition, despite the centrality of the regional concept within the field of

Geography. Some types of region are difficult to define, let alone to delimit on the ground, or to make use of in a constructive manner. One of the most successful types of region, at least in so far as it has been found useful in the real world, is the planning region. For purposes of town and country planning it is normal practice to subdivide the area under consideration into smaller units, thus achieving a simplification and facilitating the implementation of planning policies. These units may or may not be the optimum divisions, however, and planning regions are likely to be most useful if they reflect actual differences and if the differences between the regions are greater than the differences within the regions.

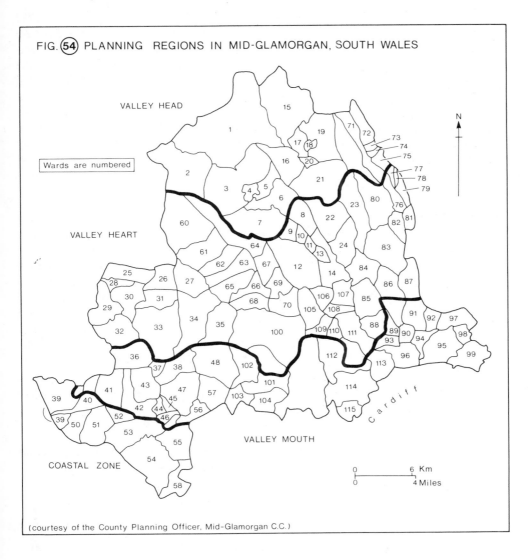

FIG. (54) PLANNING REGIONS IN MID-GLAMORGAN, SOUTH WALES

VALLEY HEAD

Wards are numbered

VALLEY HEART

VALLEY MOUTH

COASTAL ZONE

0   6 Km

0   4 Miles

N

(courtesy of the County Planning Officer, Mid-Glamorgan C.C.)

Figure 54 shows the four planning regions used by the Mid-Glamorgan County Council Planning Department in South Wales. These regions include part of the South

Wales coalfield, the Ogmore, Rhondda, Cynon and Taff valleys, and the South Wales coast and have been named as follows:

   (i) The Valley Head Region.
   (ii) The Valley Heart Region.
   (iii) The Valley Mouth Region.
   (iv) The Coastal Zone Region.

The present planning problems in the Mid-Glamorgan County are a microcosm of those of South Wales generally and, like many other areas formerly dominated by coal, owe much to the history of the development of the coalfield.

Blast furnaces began operation at Merthyr Tydfil in the Valley Head Region in 1759 and by 1820 a line of 'iron towns' had developed along the north crop of the coalfield utilizing the local resources of coal, ironstone and limestone. After 1840 iron was overshadowed by the development of coal, especially the steam coals of the centre of the coalfield, which were most accessible in the deep north–south valleys. Merthyr Tydfil remained the largest town in Wales until the late nineteenth century, however. The building boom which accompanied the development of coal mining in the Valley Heart Region was responsible for most of the housing present there today, and for the devastated landscape. The Taff–Rhondda valley system, which in the 1850s was virtually uninhabited, supported 161,000 people (almost all in coal mining) by 1911. Throughout the period of the pre-eminence of coal, the growth of Cardiff and the other ports of South Wales was phenomenal, benefiting from the export of coal and the import of iron ore for steel works, which had, by the end of the nineteenth century, moved to the coast. The Valley Mouth Region as a whole took part in the general prosperity and attracted many industries. In contrast, the period 1920–40 was one of unmitigated decline in the Valley Heart Region. The Taff–Rhondda valleys suffered the biggest decline, a result of the decline of coal mining, and although economic conditions improved after 1945, by 1966 these valleys supported only 5000 people connected with coal at the five collieries remaining. The decline of the inland areas was accompanied by continued growth of the Valley Mouth Region, which is today characterized by the rapid growth of service industries. The Coastal Zone was relatively little effected by the industrial development that accompanied the development and decline of coal mining. Today, the Coastal Zone is important for recreation and as a residential area for high-income commuters from the larger towns.

This exercise considers the four planning regions of Mid-Glamorgan with a view to answering two questions:

    (i) Are there distinct socio-economic differences between them?
and (ii) Can any differences be related to the legacy of the past?
The first question will be tested directly by use of Kruskal–Wallis tests; the second question is more difficult and will require inference on the basis of your results.

Practical work

Five socio-economic indicators for each ward within the four planning regions of Mid-Glamorgan are given in Table 16. These data were compiled from the 1971 Census returns. The indices are:

TABLE 16. *Socio-economic indicators for Mid-Glamorgan wards, South Wales*

Ward no.	Socio-economic indicator (i)	(ii)	(iii)	(iv)	(v)	Ward no.	Socio-economic indicator (i)	(ii)	(iii)	(iv)	(v)
1	0.506	0.519	0.700	0.092	0.569	60	0.344	0.619	0.475	0.087	0.514
2	0.518	0.535	0.751	0.115	0.596	61	0.375	0.825	0.499	0.102	0.572
3	0.377	0.370	0.741	0.079	0.559	62	0.388	0.806	0.506	0.080	0.632
4	0.405	0.797	0.556	0.119	0.482	63	0.386	0.737	0.448	0.061	0.566
5	0.394	0.757	0.627	0.139	0.470	64	0.314	0.698	0.521	0.063	0.593
6	0.430	0.635	0.663	0.084	0.584	65	0.366	0.776	0.360	0.069	0.553
7	0.341	0.776	0.437	0.079	0.575	66	0.395	0.704	0.459	0.062	0.620
8	0.361	0.640	0.495	0.071	0.602	67	0.345	0.530	0.573	0.032	0.520
9	0.365	0.585	0.675	0.030	0.647	68	0.387	0.630	0.563	0.084	0.562
10	0.320	0.622	0.370	0.096	0.634	69	0.403	0.661	0.445	0.107	0.494
11	0.295	0.537	0.452	0.052	0.631	70	0.387	0.539	0.539	0.044	0.635
12	0.434	0.553	0.652	0.063	0.690	71	0.386	0.408	0.739	0.066	0.618
13	0.242	0.335	0.590	0.052	0.623	72	0.331	0.360	0.813	0.063	0.550
14	0.429	0.713	0.533	0.087	0.650	73	0.418	0.798	0.503	0.106	0.575
15	0.465	0.382	0.736	0.078	0.574	74	0.344	0.489	0.584	0.035	0.474
16	0.394	0.479	0.611	0.108	0.500	75	0.363	0.380	0.633	0.071	0.548
17	0.385	0.339	0.853	0.121	0.524	76	0.344	0.678	0.275	0.032	0.508
18	0.374	0.489	0.581	0.059	0.572	77	0.395	0.675	0.388	0.044	0.652
19	0.360	0.656	0.508	0.062	0.526	78	0.251	0.141	0.139	—	0.588
20	0.417	0.603	0.650	0.133	0.509	79	0.311	0.568	0.268	0.026	0.553
21	0.382	0.643	0.455	0.062	0.517	80	0.391	0.458	0.653	0.042	0.590
22	0.330	0.474	0.454	0.066	0.615	81	0.368	0.237	0.639	0.017	0.681
23	0.397	0.585	0.558	0.085	0.525	82	0.416	0.563	0.698	0.062	0.588
24	0.405	0.681	0.501	0.106	0.578	83	0.442	0.405	0.711	0.122	0.604
25	0.336	0.629	0.611	0.060	0.676	84	0.526	0.625	0.774	0.168	0.538
26	0.367	0.643	0.405	0.019	0.705	85	0.346	0.590	0.505	0.033	0.533
27	0.364	0.747	0.366	0.044	0.707	86	0.533	0.623	0.682	0.166	0.589
28	0.369	0.809	0.453	0.071	0.452	87	0.570	0.612	0.800	0.153	0.542
29	0.431	0.588	0.720	0.149	0.552	88	0.421	0.581	0.596	0.087	0.558
30	0.433	0.691	0.674	0.067	0.674	89	0.515	0.261	0.962	0.115	0.622
31	0.390	0.786	0.442	0.097	0.597	90	0.556	0.555	0.838	0.124	0.621
32	0.490	0.501	0.821	0.135	0.678	91	0.482	0.387	0.837	0.088	0.582
33	0.417	0.259	0.770	0.034	0.625	92	0.439	0.411	0.713	0.047	0.698
34	0.398	0.481	0.608	0.108	0.641	93	0.582	0.509	0.838	0.121	0.590
35	0.350	0.416	0.578	0.044	0.622	94	0.515	0.317	0.963	0.114	0.535
36	0.593	0.419	0.767	0.071	0.643	95	0.698	0.603	0.598	0.125	0.625
37	0.506	0.587	0.831	0.138	0.630	96	0.622	0.730	0.757	—	0.571
38	0.523	0.338	0.904	0.097	0.559	97	0.576	0.436	0.924	0.198	0.603
39	0.545	0.215	0.956	0.119	0.547	98	0.590	0.748	0.870	0.263	0.589
40	0.507	0.452	0.923	0.080	0.591	99	0.851	0.617	0.787	—	—
41	0.485	0.596	0.734	0.169	0.470	100	0.514	0.486	0.792	0.142	0.539
42	0.662	0.576	0.980	0.149	0.649	101	0.529	0.457	0.772	0.117	0.683
43	0.662	0.722	0.853	0.212	0.623	102	0.624	0.585	0.836	0.194	0.583
44	0.627	0.497	0.931	0.286	0.506	103	1.000	0.833	1.000	—	—
45	0.476	0.427	0.832	0.101	0.604	104	0.484	0.122	0.893	0.015	0.594
46	0.573	0.677	0.897	0.256	0.493	105	0.449	0.783	0.529	0.126	0.541
47	0.767	0.814	0.947	0.239	0.619	106	0.501	0.401	0.921	0.089	0.580
48	0.594	0.481	0.672	0.097	0.806	107	0.383	0.711	0.585	0.082	0.577
50	0.646	0.785	0.924	0.343	0.410	108	0.487	0.801	0.627	0.164	0.617
51	0.560	0.615	0.905	0.159	0.527	109	0.337	0.661	0.365	0.067	0.548
52	0.853	0.441	0.971	—	—	110	0.422	0.767	0.551	0.159	0.561
53	0.578	0.604	0.526	0.400	0.466	111	0.408	0.239	0.845	0.067	0.496
54	0.778	0.730	0.900	0.293	0.483	112	0.725	0.712	0.932	0.188	0.623
55	0.756	0.659	0.821	0.242	0.606	113	0.510	0.427	0.830	0.106	0.553
56	0.838	0.851	0.943	0.405	0.595	114	0.692	0.600	0.915	0.253	0.544
57	0.674	0.586	0.930	0.144	0.643	115	0.830	0.283	0.774	0.333	0.333
58	0.770	0.642	0.877	0.235	0.529						

(From the County Planning Officer, Mid-Glamorgan County Council)

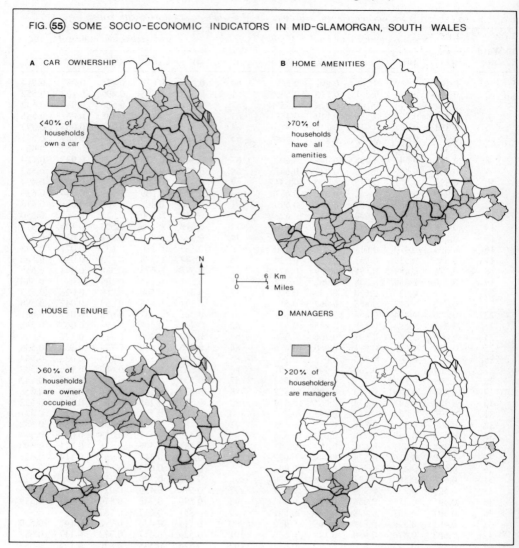

FIG. (55) SOME SOCIO-ECONOMIC INDICATORS IN MID-GLAMORGAN, SOUTH WALES

A CAR OWNERSHIP

<40% of households own a car

B HOME AMENITIES

>70% of households have all amenities

N

0 — 6 Km
0 — 4 Miles

C HOUSE TENURE

>60% of households are owner-occupied

D MANAGERS

>20% of householders are managers

 (i) *Car ownership* = the proportion of households with at least one car.
 (ii) *House tenure* = the proportion of households that is owner-occupied.
(iii) *Amenities* = the proportion of households with the exclusive use of all amenities
       (i.e. hot water, W.C., etc.).
(iv) *Managers* = the proportion of householders in Standard Economic Groups 1, 2,
       3, 4, 13 (employers and managers).
 (v) *Skilled workers* = the proportion of householders in Standard Economic Groups
       5, 6, 8, 9, 12, 14 (foremen, skilled manual workers and non-manual workers).

Each ward is numbered on Fig. 54, which shows the planning region within which it is
located. Five wards, numbered 80, 86, 100, 101 and 112, are located across the boundaries

of planning regions; these wards should not, therefore, be used in the analyses. There are no wards numbered 49 or 59. Selected aspects of four of the socio-economic indicators are mapped in Fig. 55 (A) to (D).

1. Use a Kruskal–Wallis four-sample test to test the hypothesis that there is no significant difference between the regions in terms of car ownership. All steps in the test should be clearly shown.

2. Carry out a similar test on one other of the socio-economic indicators.

3. In the light of the data, Fig. 55, and the results of the Kruskal–Wallis tests, briefly outline the main socio-economic differences between the four regions.

4. (a) Which of the following two-sample tests for independent samples could be appropriately applied to the data from the Coastal Zone and the Valley Mouth Regions to test for a difference between them:

      (i) Student's t-test?

      (ii) χ^2 test?

      (iii) Kolmogorov–Smirnov test?

      (iv) Mann–Whitney test?

  (b) Which, if any, of these tests would be most appropriate for this purpose? Give your reasons.

5. To what extent do the results of the Kruskal–Wallis test confirm that the four regions are genuine entities (cf. definition 5 of Reiner)?

6. Are the data and the results of the Kruskal–Wallis tests consistent with the view that the present socio-economic characteristics of the planning regions of Mid-Glamorgan reflect a legacy of the past?

7. This exercise has utilized a large data set but has only considered five socio-economic indicators derived from census returns. Suggest some other ways in which the regions may differ and methods that could be used to obtain suitable data for their investigation.

12

Non-parametric Tests
for Dependent (matched)
Samples

APART from the Student's t-test for two dependent samples (Chapter 9) all the tests so far considered have been appropriate for independent samples. In this chapter two non-parametric tests for dependent samples are introduced. The Wilcoxon matched-pairs signed-ranks test is the non-parametric equivalent of the Student's t-test for two dependent samples, but the Wilcoxon test can be applied to non-normal distributions and requires only ordinal-scale data. The second test to be considered is the Friedman two-way analysis of variance by ranks, which is a test for more than two dependent samples. There is great potential for more widespread use of these tests in geographical studies. To realize this potential, however, it will be necessary to adopt a more 'experimental' approach to geographical analysis.

The Wilcoxon matched-pairs signed-ranks test

Wherever the Student's t-test for dependent samples is applicable, the Wilcoxon test is applicable, although the Wilcoxon test is less powerful if the assumptions of the t-test are met. In other words, the Wilcoxon test is less likely to detect a difference in marginal cases because ordinal scale information about the differences between the samples is less precise than interval scale information. Just as the Student's t-test for dependent samples analyses the differences between each matched-pair of observations or measurements, so the Wilcoxon test analyses the rank order of the differences between the matched-pairs. The sort of situations where this test would be appropriate include: (i) measurement of changes in opinion between two time-periods (when the same group of people are interviewed on both occasions); (ii) measurement of supposedly asymmetrical valleys by cross-profiles (where slope angles have been recorded for pairs of north-and south-facing valley-side slopes); and (iii) measurement of the response of crops to two intensities of treatment by fertilizers (where the experiment involves a number of plots, and half of each plot receives treatment A and the other half of each plot receives treatment B). The following worked example is based on a set of hypothetical results from one such experiment.

Twelve plots (each with different environmental conditions of shading, stoniness, drainage, etc.) were planted with the same crop, and half of each plot was given fertilizer treatment A while the other half of each plot was given treatment B. The yields of each plot under each treatment, the difference between each pair, the rank-order of the difference between each pair (d_r) and the direction of the difference $(+ \text{ or } -)$ were as shown in Table 17.

TABLE 17

Plot	Treatment A	Treatment B	Difference	d_r	Direction of difference
		Yield (bu/acre)			
1	26.3	28.0	1.7	2	+
2	24.2	25.7	1.5	3	+
3	23.1	25.1	2.0	1	+
4	21.6	22.3	0.7	7	+
5	24.5	25.2	0.7	7	+
6	23.3	25.4	1.1	5	+
7	21.8	23.0	1.2	4	+
8	22.9	23.4	0.5	9	+
9	25.7	25.5	0.2	12	−
10	24.1	24.8	0.7	7	+
11	25.8	25.5	0.3	11	−
12	26.1	25.7	0.4	10	−

If there were in fact 'no difference' between the two samples (treatments) then the sum of the positive rankings would be expected to be about equal to the sum of the negative rankings. The greater the difference between the two samples, the smaller will be the sum of the ranks with the less frequent sign. The tabulated Wilcoxon's T statistic (Table H, Appendix) gives the smallest value of T that would be expected to occur by chance if there were indeed 'no difference' between the samples. The calculated Wilcoxon's T statistic is obtained from the sum of the ranks with the less frequent sign (R_1) and is whichever of the following two values is the smaller:

$$\text{(a) } T' = R_1 \quad \text{or} \quad \text{(b) } T'' = m(n+1) - T'$$

where $m =$ the number of ranks with the less frequent sign,

$\quad\quad\quad n =$ the sample size (number of matched-pairs).

In the example,

$$\text{(a) } T' = 33 \quad \text{and} \quad \text{(b) } T'' = 3(12+1) - 33 = 6$$

In this instance, the calculated T statistic is therefore 6.

Reference to Table H (Appendix) shows that, with a sample size of 12 and a 5% significance level, a value of T as small as 13 would be expected if there was in fact 'no difference' between the two treatments. The calculated T statistic is smaller than this; it is therefore concluded that the calculated T is smaller than can be attributed to chance and that the hypothesis of 'no difference' must be rejected. It should be noted, however, that the Wilcoxon test is the only test in the manual that requires the calculated statistic to be *smaller* than the tabulated statistic in order to *reject* the hypothesis of 'no difference' (null hypothesis).

A second example is based on population data from the thirty largest cities in Australia (Scott, 1965). The Wilcoxon test will be used to test for a significant difference between the percentage of the population that was British-born and the percentage of the population that was born on mainland Europe. The data are made up of thirty matched-pairs of measurements or observations, which are set out in Table 18.

TABLE 18

City	% of the population (1961)		Difference	d_r	Sign
	British-born	Continental Europeans			
Sydney	8.2	8.7	0.5	29	+
Melbourne	8.1	12.8	4.7	6	+
Brisbane	8.4	4.7	3.7	11	−
Adelaide	8.2	12.5	4.3	8.5	+
Perth	12.4	9.2	3.2	12	−
Newcastle	6.4	5.1	1.3	23	−
Wollongong	11.4	15.7	4.3	8.5	+
Hobart	5.9	5.1	0.8	26	−
Geelong	7.3	13.5	6.2	4.5	+
Launceston	5.9	4.0	1.9	19	−
Canberra	9.6	14.2	4.6	7	+
Ballarat	3.9	5.4	1.5	21.5	+
Townsville	5.4	2.8	2.6	16	−
Toowoomba	5.5	1.5	4.0	10	−
Latrobe Valley	12.8	13.6	0.8	26	+
Ipswich	6.2	3.4	2.8	14	−
Rockhampton	4.1	1.1	3.0	13	−
Bendigo	3.1	1.6	1.5	21.5	−
Cessnock	9.4	1.9	7.5	3	−
Gold Coast	8.9	2.7	6.2	4.5	−
Penrith	9.2	11.9	2.7	15	+
Broken Hill	1.9	4.0	2.1	18	+
Blue Mountains	11.6	3.9	7.7	2	−
Maitland	2.4	4.2	1.8	20	+
Cairns	6.1	5.2	0.9	24	−
Elizabeth	42.3	7.6	34.7	1	−
Bundaberg	4.5	2.1	2.4	17	−
Wagga Wagga	2.9	3.2	0.3	30	+
Kalgoorlie	6.9	7.5	0.6	28	+
Goulburn	3.3	2.5	0.8	26	−

The sum of the ranks with the less frequent sign is the sum of the + ranks, which is $R_1 = 222.0$.

The number of ranks with the less frequent sign is $m = 13$.

The sample size is $n = 30$.

The calculated T statistic is the smaller of:

(a) $T' = R_1 = 222$ and   (b) $T'' = 13(30 + 1) - 222 = 403 - 222 = 181$

The calculated T statistic is therefore 181.

Table H (Appendix), with a 5% significance level, shows that a value of T as low as 137 would be expected to result by chance if there was 'no difference' between the two samples. We are therefore unable to reject the hypothesis of 'no difference' (because the calculated T is not small enough). In other words, we are unable to detect a consistent difference in the relative proportions of British-born and 'Europeans' in the Australian cities.

When sample sizes are larger than those tabulated in Table H (Appendix) the significance of the calculated T can be found from tables of the normal distribution function, z being given by the formula:

$$z = \frac{T - \dfrac{n(n+1)}{4}}{\sqrt{\dfrac{n(n+1)(2n+1)}{24}}}$$

In the example,

$$z = \frac{181 - \dfrac{30(31)}{4}}{\sqrt{\dfrac{30(31)(61)}{24}}} = \frac{181 - 232.5}{\sqrt{2363.75}} = \frac{-51.5}{48.618} = -1.059$$

Table A (Appendix) shows that, using a 5% significance level ($p = 0.975$ in the table, which is arranged for a one-tailed test), a tabulated z value of 1.96 must be exceeded for rejection of the hypothesis of 'no difference'. Thus we are unable to reject the hypothesis of 'no difference'; a similar conclusion to that reached by the use of Table H.

For the limitations of the Wilcoxon test, see the Mann–Whitney test. In the case of the Wilcoxon test, however, the independent and random sampling assumptions apply within samples, not between samples (within columns, not within matched-pairs).

The Friedman two-way analysis of variance by ranks

Friedman's test is applicable to situations where there are more than two (k) dependent samples and when measurements are on at least an ordinal scale. Dependent (matched) samples may be related qualitatively with respect to some condition, state, or aspect of the environment. For example, the production of three types of paper at twelve paper mills in Kent, southern England, in 1860–65 provides three related samples with twelve individuals in each (sample size $= 12$). The rank of the three types of paper at each mill was as shown in Table 19 (Lewis, 1977).

TABLE 19

Mill	Rank order of importance of:		
	Specialist paper	Printing paper	Packing paper
Sundridge	3	2	1
Cray	3	2	1
Chafford	3	2	1
Darenth	3	1.5	1.5
Basted	2	3	1
Roughway	2	3	1
East Malling	3	2	1
Lower Tovil	3	2	1
Upper Tovil	1	3	2
Medway	1.5	1.5	3
Springfield	3	2	1
Hayle	3	2	1

The most important type of paper at each mill has been assigned rank 1. Furthermore, by ranking across a row (a matched set of three observations or measurements) differences between mills are controlled and do not influence the rankings obtained. Now, if there is no difference in the relative importance of the three types of paper throughout this area, it would be expected that the column totals would be approximately equal. The greater the difference between the column totals, the more likely it is that there exists a real difference in the relative importance of the three types of paper. The calculated χ_r^2 statistic measures the difference between the column totals, and is found from the formula:

$$\chi_r^2 = \frac{12}{nk(k+1)} \left[\sum R^2 \right] - 3n(k+1)$$

where n = the sample size (number of rows),
    k = the number of samples (number of columns),
    R = the sum of the ranks of a particular sample,
    R^2 = the square of R,
    $\sum R^2$ = the sum of the squares of the ranks of the k samples.

The formula is thus very similar to that used to calculate H in the Kruskal–Wallis test (the equivalent test for independent samples). In the example, the sum of the ranks for each sample are:

$R_1 = 30.5$ $R_2 = 26.0$ $R_3 = 15.5$

These values squared give:

930.25   676.0   240.25

$\sum R^2$ is therefore 1846.5 and the calculated χ_r^2 statistic becomes:

$$\chi_r^2 = \frac{12}{12(3)(3+1)} \, 1846.5 - 3(12)(3+1)$$

$$= \frac{12}{144} \times 1846.5 - 144$$

$$= 153.875 - 144 = 9.875$$

For very small sample sizes Table I (Appendix) gives the tabulated χ_r^2 statistic. For larger sample sizes (or for a larger number of samples) χ_r^2 approximates to χ^2, so that Table D (Appendix) can be used with $(k-1)$ degrees of freedom. In this case, using the 5% significance level, a χ_r^2 value as large as 5.99 would be expected as a result of chance. The calculated statistic is larger than the tabulated statistic; the difference between the three samples is thus greater than can be attributed to chance and the hypothesis of 'no difference' must be rejected. It is therefore concluded that there is a real difference in the relative importance of the three types of paper production in this region. A similar conclusion would probably result from a geographical interpretation of the pie-graphs in Fig. 56, but the traditional approach would not give a measure of the statistical significance of this conclusion.

For the limitations of the Friedman test, see the Mann–Whitney test. In the case of the Friedman test, however, the independent and random sampling assumptions apply within samples, not between samples (within columns, not between columns).

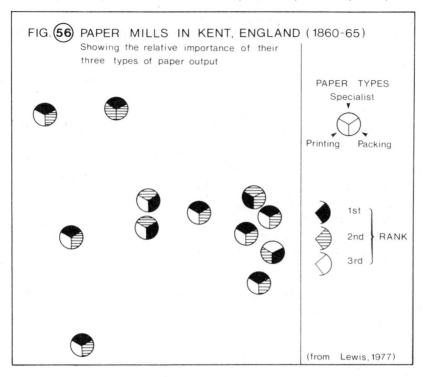

FIG. 56 PAPER MILLS IN KENT, ENGLAND (1860-65)

Showing the relative importance of their three types of paper output

PAPER TYPES

Specialist

Printing   Packing

1st
2nd } RANK
3rd

(from Lewis, 1977)

Exercise 18. Form and origin of saw-tooth moraines in Bödalen, Norway, investigated by Wilcoxon tests and Mann–Whitney tests.

Background

A series of end moraines with an unusual saw-tooth plan (Fig. 57) are found in Bödalen in front of one of the outlets of the Jostedalsbreen ice-cap. The saw-tooth pattern of the moraines appears to reflect a radial pattern of crevasses at a former glacier snout. The present-day glacier is found immediately to the south of Fig. 57, and it seems that the moraines were deposited during the 'Little Ice Age' of the last few centuries, when the glacier was more extensive than it is today. As these moraines are remarkably well preserved, and because it was thought possible that the detailed form of the moraines would reveal clues to their mode of formation, a field survey was carried out by means of cross-profiles. The location and form of the profiles are shown in Figs. 57 and 58, respectively.

The aim of the exercise is to analyse the width, height and slopes from each cross-profile, with particular reference to the similarities and differences between teeth (sections of moraine pointing down-valley) and notches (sections of moraine pointing up-valley). In this way a detailed picture of the three-dimensional form of the moraines can be built up. Precise information on the form of the moraines is a necessary prerequisite to the making of inferences about possible mechanisms for their formation.

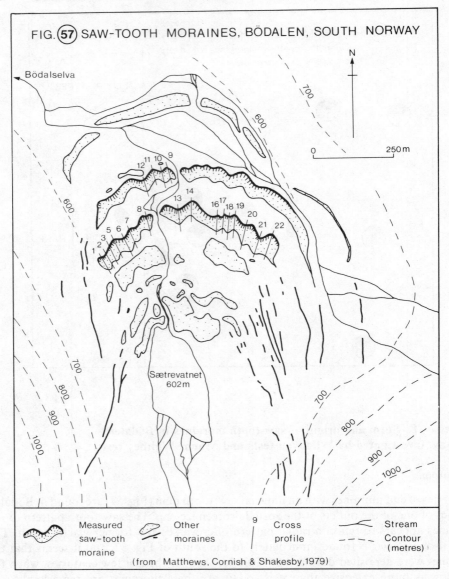

FIG. (57) SAW-TOOTH MORAINES, BÖDALEN, SOUTH NORWAY

Bödalselva

Sætrevatnet
602 m

| | Measured saw-tooth moraine | | Other moraines | 9 | Cross profile | | Stream |
| | | | | | | | Contour (metres) |

(from Matthews, Cornish & Shakesby, 1979)

The mechanisms by which end moraines are formed by glaciers is a matter of controversy. Three of the possible mechanisms can be conveniently labelled dumping, squeezing and pushing. Dumping implies that supra-glacial debris (which originates by rockfall and avalanche from the surrounding terrain) is dumped as the glacier melts. If the margin of the glacier is stationary for a sufficiently long period, then the dumped material will begin to accumulate. Squeezing implies that water-soaked, sub-glacial debris is squeezed up from beneath the glacier at the glacier snout. Such a process may form a deposit along the margin of the glacier snout. Pushing implies that, during a glacier advance, the glacier acts as a bulldozer, which results in the accumulation of pro-glacial debris in front of the advancing ice.

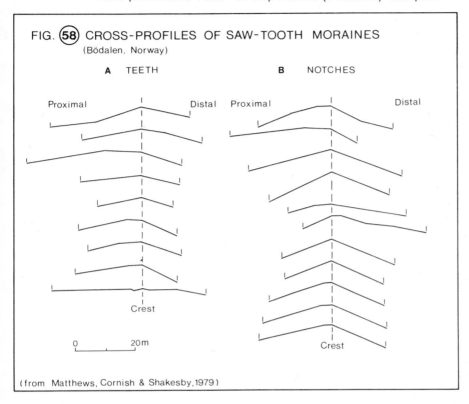

FIG. 58 CROSS-PROFILES OF SAW-TOOTH MORAINES
(Bödalen, Norway)

A   TEETH   B   NOTCHES

Proximal   Distal   Proximal   Distal

Crest

0   20m

Crest

(from Matthews, Cornish & Shakesby, 1979)

It transpires that only one of the contending mechanisms is likely to have produced the characteristic form of the saw-tooth moraines in Bödalen.

Practical work

The following data (Table 20) were derived from the profiles in Fig. 57.

TABLE 20. *Cross-profiles of saw-tooth moraines, Bödalen, southern Norway*

Teeth					Notches				
Profile no.	Width (m)	Height (m)	Proximal slope (°)	Distal slope (°)	Profile no.	Width (m)	Height (m)	Proximal slope (°)	Distal slope (°)
2	47.0	5.0	20.0	13.0	1	45.0	7.5	23.0	26.0
5	39.5	4.0	9.0	17.0	3	42.5	4.5	6.5	27.5
7	50.5	5.0	8.5	19.0	6	50.5	8.0	15.5	22.0
9	32.5	3.0	7.5	15.0	8	40.5	9.0	27.0	21.5
11	24.5	3.5	16.5	20.0	10	39.5	4.0	17.0	10.0
14	32.5	5.0	3.0	25.0	12	40.0	4.5	29.0	20.0
17	31.0	4.0	15.0	19.0	13	38.0	8.0	22.0	22.0
19	33.5	5.0	9.0	27.5	16	33.5	7.0	21.5	25.0
22	51.0	2.0	8.0	17.5	18	38.5	7.5	21.0	25.0
					20	41.5	7.5	18.0	25.0
					21	42.0	6.5	13.0	24.0

Proximal slopes are facing towards the glacier; distal slopes face away from the glacier. Measurements are made to the nearest 0.5 m and 0.5°.

Slope angles are maximum slope angles.

Width is defined as the length of the line joining the base of the proximal slope to the base of the distal slope (the base-line).

Height is defined as the vertical distance from the crest (point of maximum slope angle change) to the base-line.

1. For teeth and for notches (separately) calculate the mean of each of the following attributes:
(a) Height.
(b) Width.
(c) Proximal slope angle.
(d) Distal slope angle.

2. Briefly describe the differences in size and shape of teeth and notches as revealed by the mean values.

3. Use Mann–Whitney tests for two independent samples to throw light on the following questions:
(a) Do teeth and notches differ significantly with respect to proximal slope angles?
(b) Do teeth and notches have similar distal slope angles?
(c) Can the saw-tooth moraines be said to possess a constant width?
(d) Have saw-tooth moraines an undulating crest-line?

4. Reassess your answer to question 2 in the light of the Mann–Whitney tests.

5. Examine the asymmetry of the moraine cross-profiles, including a Wilcoxon matched-pairs signed-ranks test to compare each of the following:
(a) The proximal and distal slope angles of teeth.
(b) The proximal and distal slope angles of notches.

6. Draw a sketch (perspective view) of a short stretch of saw-tooth moraine, taking care to include all of their characteristic features.

7. Bearing in mind the analyses and the background information given previously, which of the mechanisms for end moraine formation—dumping, squeezing or pushing—accounts for all of the characteristic features of the saw-tooth moraines? Reasons for the acceptance or rejection of each mechanism should be discussed. Any further evidence that would be necessary to resolve two or more possible conclusions ought to be mentioned.

8. Can general conclusions about mechanisms of moraine formation be drawn from these unique forms?

Exercise 19. Recent trends in causes of death in some advanced Western societies characterized by application of Friedman analysis of variance by ranks.

Background

On the basis of data compiled by the United Nations and the World Health Organization (Spiegelman, 1965), the aim of the exercise is to generalize about recent changes in causes of death for a number of countries with a long history of low death rates. These include the countries of Western Europe and the U.S.A., Canada, Australia, New

Zealand and (white) South Africa. Although there are differences in the practices adopted regarding the recognition and documentation of mortality, the data can be expected to portray the overall changes in causes of death as reflected by medical opinion in these countries.

The data are given in ten tables (Table 21a–j), each of which shows the average annual number of deaths per 100,000 of the population (death rate) for each country. The first table (A) shows the number of deaths from all causes. The remaining tables cover:

    B. Cardio-vascular and renal (heart and kidney diseases)
    C. Cancer
    D. Diabetes
    E. Tuberculosis
    F. Influenza and pneumonia
    G. Cirrhosis of the liver
    H. Suicide
    I.   Motor vehicle accidents
    J. All other accidents

For each country, data are available for three periods: 1950–3, 1954–7 and 1958–61. We have, therefore, three related samples, which can be compared by means of the Friedman two-way analysis of variance by ranks for k-dependent samples. In this way it is possible to test whether or not there are consistent tendencies in the incidence of the various causes of death in this group of countries. These tendencies may in turn be related to the advancement of medical knowledge and other changes in society.

TABLE 21. *Causes of death in countries of low mortality*

(a) *Death rates from all causes*

Country	1950–3	1954–7	1958–61
U.S.A.	962	936	941
England/Wales	1172	1154	1169
Scotland	1216	1197	1209
Australia	945	899	861
New Zealand	924	910	894
Canada	884	820	788
South Africa	857	852	876*
Ireland	1267	1207	1198
Netherlands	751	758	756
Belgium	1221	1209	1181
France	1227	1212	1113
Switzerland	1017	1008	951
West Germany	1068	1101	1100
Denmark	902	899	935
Norway	863	862	905
Sweden	979	964	973
Finland	988	920	892
Portugal	1186	1145	1058
Italy	1003	971	945
Spain	1037	949	870*

(From Spiegelman, 1965.)

(b) *Death rates from cardio-vascular and renal diseases*

Country	1950–3	1954–7	1958–61
U.S.A.	488	483	488
England/Wales	560	568	571
Scotland	597	616	623
Australia	465	451	439
New Zealand	468	437	429
Canada	403	384	377
South Africa	320	330	338*
Ireland	505	550	571
Netherlands	280	311	315
Belgium	465	357	345
France	384	380	359
Switzerland	437	439	416
West Germany	364	403	407
Denmark	389	406	423
Norway	331	374	421
Sweden	443	461	473
Finland	384	402	419
Portugal	299	316	307
Italy	364	394	396
Spain	303	276	262*

(From Spiegelman, 1965.)

TABLE 21 (*cont.*)

(c) *Death rates from cancer*

Country	1950–3	1954–7	1958–61
U.S.A.	142	147	148
England/Wales	197	207	215
Scotland	195	207	213
Australia	128	130	130
New Zealand	149	146	143
Canada	128	129	129
South Africa	123	132	136*
Ireland	148	161	167
Netherlands	150	157	166
Belgium	159	208	220
France	176	185	195
Switzerland	187	190	191
West Germany	179	195	205
Denmark	174	194	209
Norway	156	160	163
Sweden	155	165	181
Finland	143	148	153
Portugal	65	82	93
Italy	115	131	146
Spain	81	99	109*

(From Spiegelman, 1965.)

(d) *Death rates from diabetes*

Country	1950–3	1954–7	1958–61
U.S.A.	16.4†	15.7	16.3
England/Wales	7.4	7.1	7.6
Scotland	9.0	9.2	10.8
Australia	12.6	12.3	11.6
New Zealand	12.2	10.8	11.8
Canada	10.9	11.0	11.5
South Africa	8.5	9.0	9.6*
Ireland	6.9	7.1	8.2
Netherlands	11.3	12.8	14.7
Belgium	18.6	24.0	23.7
France	11.2	12.0	12.1
Switzerland	14.4	13.5	13.9
West Germany	11.1	11.3	12.9
Denmark	5.2	6.3	7.3
Norway	6.6	6.8	7.9
Sweden	11.5	10.4	12.9
Finland	6.1	6.5	9.9
Portugal	5.2	6.3	6.7
Italy	9.8	11.5	12.5
Spain	6.3	7.5	8.0*

(From Spiegelman, 1965.)

(e) *Death rates from tuberculosis*

Country	1950–3	1954–7	1958–61
U.S.A.	17.7	8.9	6.2
England/Wales	28.0	13.8	8.3
Scotland	38.4	17.7	10.9
Australia	16.2	7.9	5.0
New Zealand	17.7	11.6	6.3
Canada	20.2	8.6	5.1
South Africa	17.1	8.3	7.4*
Ireland	61.8	28.2	17.5
Netherlands	14.2	6.1	3.4
Belgium	34.2	23.8	16.9
France	49.6	30.1	22.5
Switzerland	30.0	20.2	13.7
West Germany	31.6	20.0	16.0
Denmark	11.7	5.9	4.2
Norway	22.3	11.8	6.5
Sweden	19.1	10.5	7.5
Finland	69.9	39.7	27.6
Portugal	108.2	61.6	47.2
Italy	34.1	22.2	17.9
Spain	73.8	34.7	26.3*

(From Spiegelman, 1965.)

(f) *Death rates from influenza and pneumonia*

Country	1950–3	1954–7	1958–61
U.S.A.	31.3	29.2	32.8
England/Wales	61.1	55.2	67.1
Scotland	48.3	43.0	49.9
Australia	36.3	34.9	32.8
New Zealand	26.0	37.4	44.3
Canada	42.2	36.8	34.6
South Africa	57.3	53.0	57.8*
Ireland	67.3	50.3	59.2
Netherlands	35.8	28.2	26.3
Belgium	50.8	37.4	35.7
France	82.0	61.7	46.9
Switzerland	41.5	41.0	33.1
West Germany	60.7	49.9	44.8
Denmark	42.4	24.9	29.6
Norway	51.9	45.6	50.7
Sweden	40.9	47.2	47.9
Finland	56.8	52.3	37.7
Portugal	83.6	87.3	85.8
Italy	71.9	57.9	47.5
Spain	91.4	71.5	57.4*

(From Spiegelman, 1965.)

TABLE 21 (*cont.*)

(g) *Death rates from cirrhosis of the liver*

Country	1950–3	1954–7	1958–61
U.S.A.	9.9	10.6	11.1
England/Wales	2.5	2.6	2.8
Scotland	3.2	3.9	4.3
Australia	4.7	4.7	4.8
New Zealand	3.0	3.1	2.3
Canada	4.6	5.2	5.9
South Africa	7.5	6.0	6.1*
Ireland	2.0	2.1	2.2
Netherlands	2.9	3.4	3.8
Belgium	6.0†	8.2	9.1
France	21.7	30.4	28.0
Switzerland	11.6	13.4	12.4
West Germany	8.7	13.2	17.3
Denmark	5.3	7.0	8.1
Norway	2.9	3.6	3.9
Sweden	3.4	4.8	5.3
Finland	2.3	3.3	3.4
Portugal	18.7†	23.7	20.5
Italy	12.7	14.6	16.7
Spain	10.4	13.6	14.4*

(From Spiegelman, 1965.)

(h) *Death rates from suicide*

Country	1950–3	1954–7	1958–61
U.S.A.	10.5	10.0	10.6
England/Wales	10.3	11.6	11.4
Scotland	5.4	7.4	8.2
Australia	10.1	11.0	11.5
New Zealand	9.8	9.1	9.1
Canada	7.4	7.3	7.5
South Africa	10.2	11.4	13.0*
Ireland	2.4	2.4	2.9
Netherlands	6.1	6.2	6.8
Belgium	13.3	14.2	14.3
France	15.3	16.5	16.3
Switzerland	22.0	21.7	19.4
West Germany	18.4	18.9	18.8
Denmark	23.5	22.8	19.9
Norway	7.1	7.4	7.0
Sweden	16.6	18.7	17.4
Finland	16.6	20.8	20.6
Portugal	10.0‡	9.1§	8.8
Italy	6.6	6.6	6.2
Spain	5.9	5.5	5.2

(From Spiegelman, 1965.)

(i) *Death rates from motor vehicle accidents*

Country	1950–3	1954–7	1958–61
U.S.A.	23.9	23.0	21.1
England/Wales	9.8	10.9	13.6
Scotland	9.8	10.9	12.6
Australia	23.3	23.6	24.4
New Zealand	13.6	16.2	16.3
Canada	19.3	20.7	20.9
South Africa	17.7	20.4	27.5*
Ireland	5.8	7.5	9.0
Netherlands	9.4	13.8	15.5
Belgium	11.5‖	12.5	17.5
France	9.7	18.6	18.7
Switzerland	14.8	19.1	21.9
West Germany	15.5	23.0	24.4
Denmark	9.9	14.8	17.0
Norway	5.0	7.5	8.9
Sweden	10.5	13.0	14.3
Finland	8.3†	11.0	15.3
Portugal	19.1**	6.9§	9.3
Italy	10.3	16.1	17.6
Spain	2.9	5.3	6.7*

(From Spiegelman, 1965.)

(j) *Death rates from all other accidents*

Country	1950–3	1954–7	1958–61
U.S.A.	37.4	33.4	30.6
England/Wales	23.3	24.9	24.7
Scotland	35.1	34.7	34.0
Australia	33.2	31.6	27.6
New Zealand	29.0	32.2	30.2
Canada	38.3	36.2	32.4
South Africa	30.5	29.4	30.2
Ireland	22.5	23.6	22.2
Netherlands	25.1	21.2	21.1
Belgium	29.9†	40.2	36.0
France	47.5	42.7	41.5
Switzerland	40.7	37.6	38.1
West Germany	33.7	34.2	31.7
Denmark	31.8	29.3	28.3
Norway	38.6	38.3	37.2
Sweden	28.1	26.8	28.5
Finland	47.1‖	38.9	36.2
Portugal	–	32.3§	28.1
Italy	22.1	21.3	23.2
Spain	24.3	22.5	21.3*

(From Spiegelman, 1965.)

* = 1958–60	‡ = 1950–51	‖ = 1951–53
† = 1952–53	§ = 1955–57	** = 1952

Practical work

1. (a) Using the data on deaths from all causes (A), carry out a Friedman test on the three dependent samples, giving a full account of all stages of the test.
   (b) What was the hypothesis of 'no difference' in the test?
   (c) Assuming that the test resulted in rejection of the hypothesis of 'no difference', which of the following is indicated:
      (i) The countries show a similar tendency to an increase in death rate over time.
      (ii) The countries show no significant increase in death rate over time.
      (iii) The countries show no significant increase or decrease in death rate over time.
      (iv) The countries show similar patterns of change in death rate over time.
      (v) The countries show no similarity in their patterns of change over time.
2. (a) Investigate the recent trends in the nine individual causes of death using Friedman tests. Up to nine tests can be made.
   (b) Write a short comparative account of your results, summarizing the changes in the causes of death. Limit your discussion to what can be concluded from the tests themselves.
3. The overall change in death rates between 1950–53 and 1958–61 is summarized for the same countries in Table 22.

TABLE 22. *Summary table of causes of death in countries of low mortality*

| Country | % change in death rate 1950–53 to 1958–61 | | | | | | | | |
	B	C	D	E	F	G	H	I	J
U.S.A.	0.0	4.2	−0.6*	−65.0	4.8	12.1	1.0	−11.7	−18.2
England/Wales	2.0	9.1	2.7	−70.4	9.8	12.0	10.7	38.8	6.0
Scotland	4.4	9.2	20.0	−71.6	3.3	34.4	51.9	28.6	−3.1
Australia	−5.6	1.6	−7.9	−69.1	−9.6	2.1	13.9	4.7	−16.9
New Zealand	−8.3	−4.0	−3.3	−64.4	70.4	−23.3	−7.1	19.9	4.1
Canada	−6.5	0.8	5.5	−74.8	−18.0	28.3	1.4	8.3	−15.4
South Africa	5.6	10.6	12.9	−56.7	0.9	−18.7	27.5	55.4	−1.0
Ireland	13.1	12.8	18.8	−71.7	−12.0	10.0	20.8	55.2	−1.3
Netherlands	12.5	10.7	30.1	−76.1	−26.5	31.0	11.5	64.9	−15.9
Belgium	−25.8	38.4	27.4	−50.6	−29.7	51.7	7.5	52.2	20.4
France	−6.5	10.8	8.0	−54.6	−42.8	29.0	6.5	92.8	−12.6
Switzerland	−4.8	2.1	−3.5	−54.3	−20.2	6.9	−11.8	48.0	−6.4
West Germany	11.8	14.5	16.2	−49.4	−26.2	98.9	2.2	57.4	−5.9
Denmark	8.7	20.1	40.4	−64.1	−30.2	52.8	−15.3	71.7	−11.0
Norway	27.2	4.5	19.7	−70.9	−2.3	34.5	−1.4	78.0	−3.6
Sweden	6.8	16.8	12.2	−60.7	17.1	55.9	4.8	36.2	1.4
Finland	9.1	7.0	62.3	−60.5	−33.6	47.8	24.1	84.3	−23.1
Portugal	2.7	43.1	28.8	−56.4	2.6	9.6	−12.0	−51.3	−13.0†
Italy	8.8	27.0	27.6	−47.5	−33.9	31.5	−6.1	70.9	5.0
Spain	−13.5	34.6	−27.0	−64.4	−37.2	38.5	−11.9	131.0	−12.3

(From Spiegelman, 1965.)

* = 1952–53 to 1958–61   † = 1954–57 to 1958–61.

(a) Analyse these data using a Friedman test.
(b) What conclusions are possible from the application of a Friedman test to Table

22 and why do possible conclusions differ from those obtained from the previous questions?

4. (a) In the light of your geographical and general knowledge, suggest some explanations for the changes that have occurred in the incidence of the various causes of death since 1950.

  (b) What future changes might be expected in these causes of death in these countries:

    (i) in the short-term?

    (ii) in the long-term?

13

The Strength of Relationships: Correlation Coefficients

MANY hypotheses of interest to geographers involve not only questions about overall differences and similarities between data sets but also the degree to which one data set is reflected in, associated with, related to or correlated with variability in another data set. We might show, for example, that within a region of tropical Africa, areas with different quantities of soil nutrients differ significantly in the population that they support. One of the tests previously outlined could be used for this purpose. Such a test does not, however, enable us to say to what degree the variability in soil quality is associated with variability in population. Measures of association or correlation are available for this type of problem, which can be described as a problem involving the strength of relationship between two measurable attributes or variables. The problem becomes one of statistical inference if we wish to be sure that a given strength of relationship differs significantly from 'no relationship' or from a relationship likely to have occurred by chance.

Figure 59 consists of scatter graphs representing ('perfect', 'strong', 'weak' and 'no' relationship between two variables. In each graph, a point represents an observation or measurement of two variables, x and y; the axes are the respective measurement scales. In Fig. 59 A, the value of x is always high when the value of y is high, and low values of one variable are always associated with low values of the second variable. This is perfect *positive correlation*. A perfect *negative correlation*, where a high value of one variable is associated with a low value of the other variable, is shown in Fig. 59 B. Weaker relationships are shown in Fig. 59 C to E, where the scatter of points indicates that the variability of one variable is associated with the variability of the other variable, but to a lesser degree than in Fig. 59 A and B. Extreme cases of 'no relationship' are shown in Fig. 59 F to H. A *correlation coefficient* (described in the following section) can be viewed as a measure of the extent to which the relationship between two variables departs from the situations shown in Fig. 59 F to H, and approaches the situations shown in Fig. 59 A and B.

Pearson's correlation coefficient (r)

Pearson's correlation coefficient, also known as the product-moment correlation coefficient, is a ratio of the extent to which two variables vary together to the overall

138

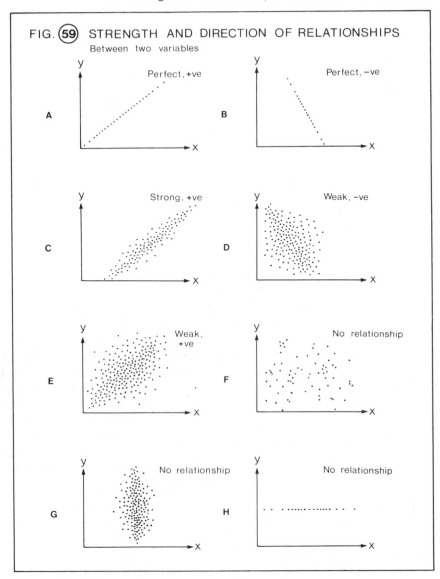

FIG. ⑤⑨ STRENGTH AND DIRECTION OF RELATIONSHIPS
Between two variables

A — Perfect, +ve

B — Perfect, –ve

C — Strong, +ve

D — Weak, –ve

E — Weak, +ve

F — No relationship

G — No relationship

H — No relationship

variability in the two sets of data. Mathematically, this is expressed as the *ratio of the covariance to the product of the standard deviations* of the two variables:

$$r = \frac{\dfrac{\sum (x - \bar{x})(y - \bar{y})}{n}}{s_x \cdot s_y} = \frac{\text{Covariance of } x \text{ and } y}{\text{Product of the standard deviation of } x \text{ and the standard deviation of } y}.$$

It is clear from the above formula that as the proportion of the variability that is involved in the covariance increases, so the correlation coefficient increases. Figure 60 can

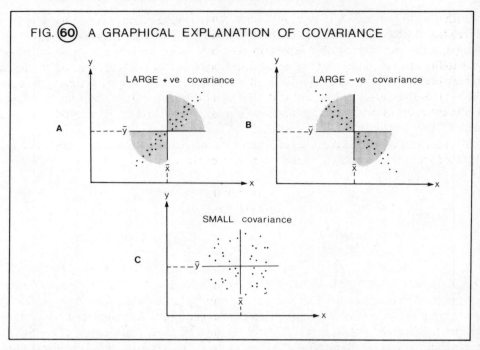

FIG. 60 A GRAPHICAL EXPLANATION OF COVARIANCE

be used to explain why this is so, and to explain the meaning of *covariance*. In Fig. 60 the means of two variables, $\bar{x}$ and $\bar{y}$, are shown superimposed on each of the scatter graphs. Covariance depends on the difference between each value of x and the mean value of x, in relation to the corresponding difference between each value of y and the mean value of y. Consider Fig. 60 A, in which all points lie in the shaded quadrants. In this case $(x - \bar{x})$ is positive when $(y - \bar{y})$ is positive, and they are also negative together (represented by the upper right and lower left quadrants, respectively). Consequently $(x - \bar{x})(y - \bar{y})$ is positive for all points, covariance is large and positive, and the correlation coefficient is large and positive. Similarly, in Fig. 60 B, $(x - \bar{x})(y - \bar{y})$ is large and covariance is large. However, in this case, $(x - \bar{x})(y - \bar{y})$ is negative for all points, so that covariance and the correlation coefficient are large and negative.

Where only a weak relationship exists between the two variables (Fig. 60 C), the points are scattered throughout the four quadrants; some of the values for $(x - \bar{x})(y - \bar{y})$ are therefore positive and some are negative; covariance is the summation of these values and is therefore small, and the correlation coefficient is small. No relationship implies an equal distribution of points between the four quadrants, a covariance of zero and a correlation coefficient of zero.

The calculated value of Pearson's r can vary between $+1.0$ (perfect positive correlation) through zero (no relationship) to -1.0 (perfect negative correlation). The involvement of n (sample size) in the calculation of covariance, and the standardization with reference to the overall variability of the data when r is calculated as a ratio, ensure this convenient and easily interpreted range of values. The greater the calculated correlation coefficient, the less likely it is to differ from zero as a result of chance. The value of r that is likely to occur by chance at a given probability level, and a given sample size, can be found in statistical tables. Thus to reject a hypothesis of 'no difference' from zero, the calculated r must exceed the tabulated r (using $n - 2$ degrees of freedom).

With a small sample size (few points in Fig. 60) it is quite possible for a relatively large correlation coefficient to result by chance. For example, if a small number of points were located, using random number tables, in relation to two axes, there would be a high probability of a reasonably strong correlation coefficient. A given correlation coefficient is therefore not as 'significant' with a small sample size as with a large sample size. It also follows that to be able to claim a statistically significant correlation a smaller correlation coefficient is necessary as sample size becomes larger, because the likelihood of a given value of r occurring by chance is reduced.

In order to avoid the necessity for a separate table of r values, a calculated t can be obtained from the calculated r value by means of the formula:

$$t = \frac{r}{\hat{\sigma}_r} = \frac{\text{Correlation coefficient}}{\text{Standard error of the correlation coefficient}}$$

$$= r \cdot \sqrt{\frac{n-2}{1-r^2}}.$$

The tabulated t statistic is then looked-up in the usual way with $(n-2)$ degrees of freedom.

A worked example will be based on the data given in Fig. 61, which shows the relationship between the pH (acidity) of certain lakes according to their distance from the Copper Cliff nickel smelter near Sudbury, Ontario, Canada. Inspection of the scatter graph suggests a weak positive correlation between the two attributes (variables) of the

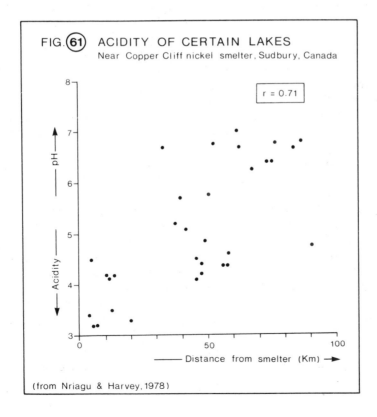

FIG. (61) ACIDITY OF CERTAIN LAKES
Near Copper Cliff nickel smelter, Sudbury, Canada

r = 0.71

pH →

Acidity →

Distance from smelter (Km) →

(from Nriagu & Harvey, 1978)

lakes. Use of Pearson's correlation coefficient enables measurement of:
      (i) the strength of the relationship
and   (ii) whether the correlation is significantly different from zero. The coefficient will
be calculated using the following formula, which is suitable for more rapid calculation
than the formula provided above but gives identical results:

$$r = \frac{n(\Sigma xy) - (\Sigma x)(\Sigma y)}{\sqrt{[n(\Sigma x^2) - (\Sigma x)^2] \cdot [n(\Sigma y^2) - (\Sigma y)^2]}}$$

where n = the sample size,

    Σx = sum of the individual values of variable x,

    Σy = sum of the individual values of variable y,

    $(\Sigma x)^2$ = sum the individual values of variable x and square the total,

    $(\Sigma y)^2$ = sum the individual values of variable y and square the total,

    (Σx^2) = square the individual values of variable x and sum the squares,

    (Σy^2) = square the individual values of variable y and sum the squares,

    (Σxy) = sum of the product of each pair of x and y values.

The data for thirty-two lakes are as shown in Table 23.

TABLE 23

Lake	Distance to smelter (km)	Acidity (pH)	Lake	Distance to smelter (km)	Acidity (pH)
Hannah	3.9	3.40	Panache	32.9	6.70
St. Chalres	4.5	4.53	Carlyle	49.0	4.85
'e'	5.2	3.20	Kakakise	50.3	5.75
Silver	6.5	3.20	Lang	52.3	6.75
Lohi	10.3	4.20	Acid (Lum II)	57.4	4.39
Raft	11.6	4.15	Lumsden I	58.1	4.39
Clearwater	12.9	3.50	Lumsden III	58.1	4.60
Tilton	13.5	4.20	Apsey	61.3	7.01
Wavy	20.0	3.30	Frood	62.6	6.70
Broker	37.4	5.20	Grab	67.0	6.25
Tyson	39.4	5.70	Evangeline	73.6	6.42
Log Boom	41.9	5.19	Maple	75.5	6.40
Johnnie	45.2	4.15	Cutler	76.7	6.79
Ruth-Roy	45.2	4.50	La Cloche	83.9	6.68
Norway	47.1	4.20	Little La Clo.	85.8	6.80
Perdix	47.7	4.41	Owl	90.3	4.75

(From Nriagu and Harvey, 1978.)

The required values for substitution in the formula are:

$$n = 32$$
$$\Sigma x = 1427.1$$
$$\Sigma y = 162.26$$

$$(\Sigma x)^2 = 2{,}036{,}614.4$$
$$(\Sigma y)^2 = 26{,}328.307$$
$$(\Sigma x^2) = 84{,}895.83$$
$$(\Sigma y^2) = 870.7698$$
$$(\Sigma xy) = 7957.853$$

The correlation coefficient is therefore:

$$r = \frac{32(7957.853) - (1427.1)(162.26)}{\sqrt{[32(84{,}895.83) - 2{,}036{,}614.4] \times [32(870.7698) - 26{,}328.307]}}$$

$$= \frac{(254{,}651.29 - 231{,}561.24)}{\sqrt{(2{,}716{,}666.5 - 2{,}036{,}614.4) \times (27{,}864.633 - 26{,}328.307)}}$$

$$= \frac{23{,}090.05}{\sqrt{(680{,}052.1)(1536.326)}} = \frac{23{,}090.05}{32{,}323.056} = 0.714.$$

The calculated Pearson coefficient is thus $+0.714$, and its statistical significance is obtained by use of the calculated Student's t statistic:

$$t = r \cdot \sqrt{\frac{n-2}{1-r^2}}$$

$$= 0.714\,35 \times \sqrt{\frac{30}{1 - 0.510\,296}} = (0.714\,35)(7.826\,97) = 5.5912$$

Reference to Table C (Appendix), and using a 5.0% significance level, shows that a calculated t statistic as large as 2.02 would be likely if there was 'no relationship' between the two variables (with $n-2 = 30$ degrees of freedom). The calculated t statistic (and hence the correlation coefficient) is therefore greater than is likely to be the result of chance, and we reject the hypothesis of 'no relationship'. In other words there is a less than 5% chance that the two variables are unrelated and we are 95% certain of a relationship. The test does *not* indicate that we are 95% sure of a correlation coefficient of $+0.714$, only that we are 95% sure of the correlation coefficient being different from zero. A statistically significant positive correlation coefficient indicates that there is a tendency for pH to increase with increasing distance from the smelter. It appears, therefore, that atmospheric pollution from the smelter provides the explanation for the low pH (high acidity) of lakes near the smelter and that there is a decreasing effect with distance. It must be emphasized, however, that inferences such as these do not follow from, but are additional to, the statistical analysis. Similarly, a correlation coefficient does not tell us which of the two variables (if any) influences the other. Any inferences about the direction of causation must be based on something more than a correlation coefficient.

Correlation coefficients are particularly useful for the comparison of maps, and for describing the degree of correspondence between distribution patterns in space. Figure 62A and B show two variables – rural farm population and mean annual precipitation – in Nebraska, U.S.A. High values of precipitation appear to be associated with high densities of the farm population. Robinson and Bryson (1957) used these maps to show how a correlation coefficient can be used to give a precise measure of the degree to which the two

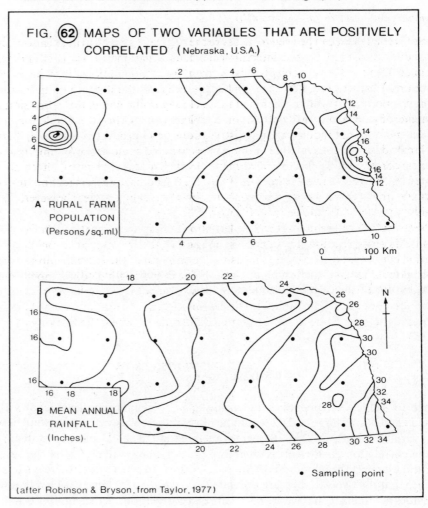

FIG. ⓺ MAPS OF TWO VARIABLES THAT ARE POSITIVELY
CORRELATED (Nebraska, U.S.A.)

A RURAL FARM
   POPULATION
   (Persons / sq. ml.)

0   100 Km

B MEAN ANNUAL
   RAINFALL
   (Inches)

N

• Sampling point

(after Robinson & Bryson, from Taylor, 1977)

maps are related. A sample of points from the map supplied the data for calculation of
Pearson's coefficient, which indicates a moderate positive correlation (+ 0.8). Once again,
correlation does not imply causation, so that it would be invalid to suggest (on the basis of
the correlation coefficient alone) that high rainfall causes high rural farm population
densities in Nebraska. It would be equally invalid to suggest, on the basis of the correlation
coefficient, that high population densities cause high rainfall!

Being a parametric statistical technique, the use of the correlation coefficient (r) as an
inferential statistic has some important limitations. Particularly important is the
requirement of interval-scale data and the necessity for an underlying bi-variate normal
distribution. A bi-variate normal distribution has each variable normally distributed
individually, and in relation to each other. The technique also assumes a linear relationship
between the two variables, and that the individuals measured are an independent random
sample.

Spearman's rank correlation coefficient (r_s)

Pearson's coefficient can be used to measure the degree of statistical association between two variables measured on the interval scale. It is a parametric statistic and tests of significance require the variables to be drawn from populations that have normal distributions. Furthermore, Pearson's coefficient measures the strength of linear relationships only, a point that will be examined more closely in the context of regression in the next chapter. Spearman's rank correlation coefficient (r_s or rho) is suitable for ordinal-scale data and does not require normal distributions or a linear relationship. If, therefore, interval scale data are not available or one wishes to avoid some of the assumptions of the parametric correlation coefficient, Spearman's coefficient is a very useful alternative. The calculated r_s statistic varies from -1.0 to $+1.0$ and is interpreted and tested for significance in the same way as Pearson's r. For small sample sizes, however, Table J, (Appendix), gives the tabulated r_s statistic.

The distribution of the coloured population of Britain is summarized in Fig. 63. The pattern bears some relationship to the distribution of the white population. Spearman's correlation coefficient can be used to measure the strength of the relationship between the coloured population and the white population. The centres with coloured populations of over 5000 in 1971 will be used here. These centres are arranged in rank order according to their coloured population in Table 24. The centres are also given a rank according to their white population, and the table gives the difference between the rankings for each centre (d).

TABLE 24

Centre	Coloured population	Rank according to coloured population	Rank according to white population	Difference in ranks (d)	d^2
Greater London	547,588	1	1	0	0
Birmingham	92,632	2	2	0	0
Wolverhampton	28,853	3	11	+8	64
Leicester	27,826	4	10	+6	36
Bradford	26,195	5	9	+4	16
Manchester	22,484	6	3	−3	9
Coventry	19,968	7	7	0	0
Leeds	16,938	8	5	−3	9
Nottingham	15,017	9	8	−1	1
Warley	13,433	10	15	+5	25
Huddersfield	12,132	11	13	+2	4
Walsall	11,956	12	19	+7	49
Luton	10,694	13	16	+3	9
Sheffield	10,551	14	4	−10	100
Derby	10,296	15	12	−3	9
Slough	10,010	16	23	+7	49
Bristol	9499	17	6	−11	121
West Bromwich	8722	18	14	−4	16
Bolton	8346	19	17	−2	4
Reading	6586	20	18	−2	4
Blackburn	6313	21	20	−1	1
Preston	6112	22	21	−1	1
Bedford	5891	23	24	+1	1
High Wycombe	5110	24	25	+1	1
Rochdale	5011	25	22	−3	9

(Data from Jones, 1978.)

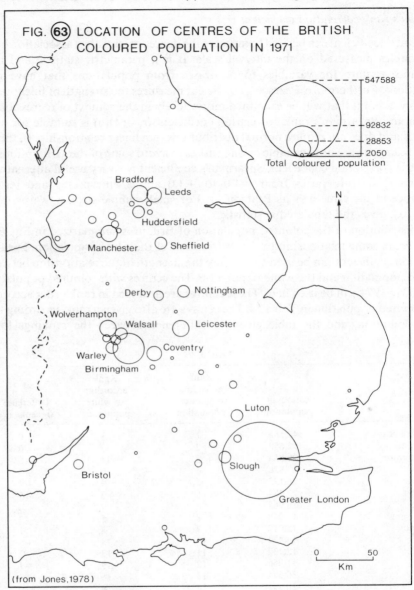

FIG. (63) LOCATION OF CENTRES OF THE BRITISH COLOURED POPULATION IN 1971

Total coloured population

547588
92832
28853
2050

Bradford
Leeds
Huddersfield
Manchester   Sheffield
Derby   Nottingham
Wolverhampton
Walsall   Leicester
Coventry
Warley
Birmingham
Luton
Slough
Bristol
Greater London

N

0   50
Km

(from Jones, 1978)

A perfect positive correlation (identical rankings in the two columns) would give rise to zero's in the right-hand column, whereas a perfect negative correlation (maximum disagreement between rankings) would give rise to very large values for the last columns. Consequently, the sum of the squares of the differences in the rankings (Σd^2) is very small or very large when the correlation is strong. The formula for Spearman's correlation coefficient is:

$$r_s = 1 - \frac{6\Sigma d^2}{n^3 - n}$$

where n = sample size,

Σd^2 = the sum of the squares of the differences between rankings.

When Σd^2 is very small, r_s is large and positive; when Σd^2 is very large, r_s is large and negative. Whether r_s is sufficiently large (positive or negative) to constitute a significant relationship, requires consideration of Table J (Appendix).

In the example, $\Sigma d^2 = 538$ and r_s becomes:

$$r_s = 1 - \frac{6(538)}{15,625 - 25}$$

$$= 1 - \frac{3228}{15,600} = 1 - 0.2069 = +0.71.$$

Table J (Appendix) indicates that using a 5% significance level and a sample size of $n = 25$, a calculated r_s value as large as 0.362 (positive or negative) would be expected if there was 'no relationship' between the two variables. We therefore reject the hypothesis of 'no relationship' and conclude that the distribution of the coloured population is related to the distribution of the white population.

For larger sample sizes than those tabulated in Table J (Appendix), r_s can be used to calculate a Student's t statistic, using the formula given in connection with Pearson's r. Spearman's r_s is simply substituted for Pearson's r in that formula; the tabulated Student's t statistic is then looked-up in the usual way. The tabulated r_s statistic (like the tabulated t statistic) is arranged for two-tailed tests but one-tailed tests can be applied by the use of the appropriate column. A one-tailed test would be appropriate if it was required to test for a significant positive correlation (or for a significant negative correlation) rather than for a significant correlation (positive or negative).

The limitations of Spearman's rank correlation coefficient are relatively few:

1. At least ordinal scale data are required.
2. If the number of tied ranks is large, then r_s may be affected; however, the effect of ties is usually negligible.
3. Independent random sampling is assumed.
4. Although r_s does not assume a linear relationship between the two variables (unlike Pearson's coefficient), it does assume a monotone relationship; that is, it indicates the strength and direction of a rising or falling relationship. A linear relationship is a more stringent requirement, implying rising or falling values in a particular (linear) form.

The contingency coefficient (C)

The contingency coefficient is calculated from χ^2 and is suitable for measuring the degree of association between two (or more) nominal-scale variables. C has a value of zero when there is 'no relationship' but does not reach 1.0 when there is a perfect relationship. The calculated value is influenced by the dimensions of the contingency table on which it is based, so that two or more contingency coefficients are only comparable if derived from contingency tables with the same number of cells.

The contingency coefficient is given by the formula:

$$C = \sqrt{\frac{\chi^2}{n + \chi^2}}$$

where n = sample size (total number of frequencies in the contingency table).

The example is taken from the results of Exercise 10, which have been arranged in the form of a 2×2 contingency table:

Soil types	Parent materials	
	Alluvium	Till
Gleyed soil	15	5
Brown earths	5	30

We want to know the degree to which the soil types are associated with particular parent materials. χ^2 is calculated in the usual way and is found to be 20.274. The contingency coefficient is therefore:

$$C = \sqrt{\frac{20.274}{55 + 20.274}} = \sqrt{\frac{20.274}{75.274}} = \sqrt{0.2693} = 0.519.$$

A test for the significance of χ^2 is also a test of the hypothesis that the contingency coefficient is 'no different' from zero. The tabulated χ^2 statistic, using a 5% significance level and $(h-1)(k-1) = (1)(1) = 1$ degree of freedom, is found to be 3.84. The calculated χ^2 exceeds the value that is necessary for the contingency coefficient to be regarded as significant. The strength of the relationship between soil type and parent material is thus $C = 0.519$, a value that is significantly different from zero.

Exercise 20: An investigation into the processes forming modern beach ridges on Jura, Scottish Inner Hebrides, using Pearson's correlation coefficient

Background

Beach ridges, composed of small cobbles 3–10 cm in diameter or boulders up to 50 cm in diameter, are the most common depositional coastal landforms on the west coast of Jura and neighbouring islands off the west coast of Scotland. The beach ridges are found in the backshore zone, above the high-water mark of ordinary spring tides (H.W.M.O.S.T.) (Fig. 64).

A levelling survey of the altitudes of forty-eight ridge crests indicated altitudes ranging from 2.8–7.7 m O.D. (metres above Ordnance Datum) with a mean altitude of 4.6 m. It has been proposed that the beach ridges are formed during high wave energy storm conditions. The aim of the exercise is to analyse the regional variation of beach ridge altitudes in an attempt to substantiate the proposal that the beach ridges are storm ridges.

Figure 65 shows the location of the beach ridges and also the location of measurement sites of the '*Pelvetia* line'. The latter feature is the landward limit of the dark-brown seaweed, *Pelvetia canaliculatus*, which is deposited at H.W.M.O.S.T. A levelling survey of the '*Pelvetia* line' during May–September 1977, when no storms occurred, showed a range of altitudes from 1.9 m to 3.1 m and a mean altitude of 2.43 m. The '*Pelvetia* line' for 1977 is thus a good indicator of the altitude limit of low-energy wave activity on this coast.

The problem is approached by the application of Pearson's correlation coefficient to relationships between the altitude of the '*Pelvetia* line' and beach ridges, and certain factors that influence wave activity under low-energy or high-energy conditions. The factors

TABLE 25. *'Pelvetia line' and beach ridge altitudes on Jura, Scottish Inner Hebrides*

Site no.	'Pelvetia line' altitude (m)	Width of the inter tidal zone (m)	Site no.	Beach ridge altitude (m)	Width of the inter tidal zone (m)	Angle of open Atlantic fetch (°)
1	2.75	7	52	3.54	30	11
2	2.70	7	53	3.26	25	2
3	2.20	20	54	2.88	32	1*
4	2.06	26	55	4.36	15	27
5	2.05	50	56	3.41	10	21
6	2.18	45	57	3.70	15	8
7	2.71	8	58	4.66	30	26
8	2.46	23	59	3.63	63	2
9	1.97	40	60	2.85	55	1
10	2.47	30	61	4.30	35	17
11	2.40	21	62	4.81	45	8
12	2.03	35	63	5.14	53	33
13	2.64	14	64	7.31	42	29
14	2.81	10	65	3.54	17	16
15	2.02	45	66	4.46	30	25
16	1.99	50	67	3.78	20	1
17	2.42	11	68	3.91	23	23
18	2.93	15	69	3.81	30	24
19	2.43	50	70	6.53	44	20
20	2.16	26	71	4.12	50	14
21	2.39	11	72	4.21	30	27
22	2.82	12	73	3.04	39	24
23	2.58	10	74	3.71	28	25
24	2.32	15	75	4.79	13	19
25	2.12	55	76	4.77	8	1
26	2.11	37	77	5.03	14	26
27	2.05	40	78	6.55	9	26
28	2.61	25	79	3.66	35	3
29	2.61	32	80	4.49	16	7
30	2.82	15	81	4.76	21	4
31	2.65	30	82	5.97	7	8
32	2.22	45	83	4.65	15	7
33	2.74	25	84	3.26	8	3
34	2.10	40	85	3.71	37	1
35	2.72	14	86	4.22	15	3
36	2.59	20	87	5.24	11	28
37	2.17	55	88	4.39	40	29
38	2.90	10	89	4.26	15	1
39	2.47	37	90	5.32	33	30
40	2.52	45	91	4.54	14	30
41	2.82	17	92	5.44	11	28
42	2.50	30	93	4.98	19	23
43	2.40	22	94	5.32	25	30
44	2.25	28	95	7.67	18	30
45	2.36	55	96	5.32	115	38
46	2.18	45	97	5.00	140	46
47	2.26	23	98	5.91	120	33
48	2.30	38	99	7.23	105	40
49	2.05	55				
50	3.07	5				
51	3.07	8		(* 0° is entered as 1°)		

(After Dawson, pers. comm.)

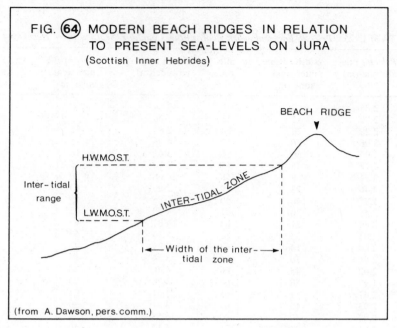

FIG. ⑥④ MODERN BEACH RIDGES IN RELATION
TO PRESENT SEA-LEVELS ON JURA
(Scottish Inner Hebrides)

BEACH RIDGE

H.W.M.O.S.T.

Inter–tidal
range

INTER-TIDAL ZONE

L.W.M.O.S.T.

Width of the inter–
tidal zone

(from A. Dawson, pers. comm.)

investigated are the width of the intertidal zone (Fig. 64) and the angle of open Atlantic
fetch (Fig. 66). Although, as has been emphasized, a correlation does not necessarily
indicate a causal relationship, the exercise is based on the principle that if a causal link
exists between two variables then one would expect a strong correlation between those
variables.

Practical work

1. Using the data given in Table 25, plot the following relationships as scatter-graphs:
(a) '*Pelvetia* line' altitude against width of the intertidal zone.
(b) Beach ridge altitude (ridge crests were measured) against width of the intertidal
zone.
(c) Beach ridge altitude against angle of open Atlantic fetch.
2. For each of the scatter-graphs, indicate:
(a) whether or not there appears to be a correlation between the two variables;
(b) whether any correlation appears to be positive or negative.
3. Calculate Pearson's correlation coefficient for each of the three relationships and test
the statistical significance of the relationships. All steps should be shown and results
should be clearly stated in terms of probabilities.
4. Are the correlation coefficients involving the *width of the intertidal zone* consistent
with the following suggestions:
(a) the waves responsible for the '*Pelvetia* line' were influenced by the width of the
intertidal zone?;
(b) the waves responsible for the deposition of the beach ridges were influenced by the
width of the intertidal zone?;

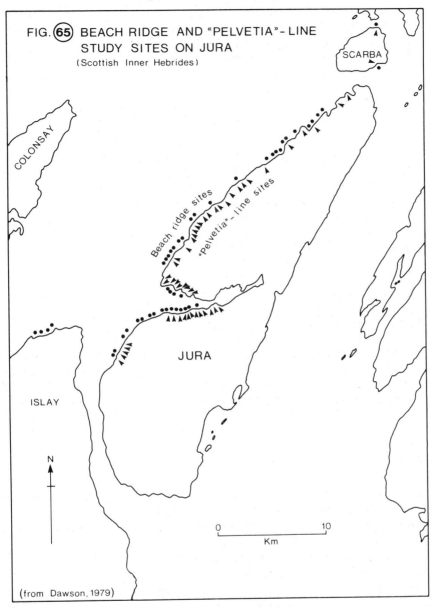

FIG. ⑥⑤ BEACH RIDGE AND "PELVETIA"-LINE
STUDY SITES ON JURA
(Scottish Inner Hebrides)

SCARBA

COLONSAY

Beach ridge sites

"Pelvetia"-line sites

JURA

ISLAY

N

0   10
Km

(from Dawson, 1979)

(c) the waves responsible for the deposition of the beach ridges were storm waves?
Fully explain your answer.

5. To what extent does the correlation coefficient between the altitude of the beach ridges and the *angle of open Atlantic fetch* support the proposal that the beach ridges are storm ridges?

6. (a) Suggest some additional factors that may account for some of the variation in the altitude of the beach ridges.

(b) How might the storm-wave origin of the beach ridges be tested further?

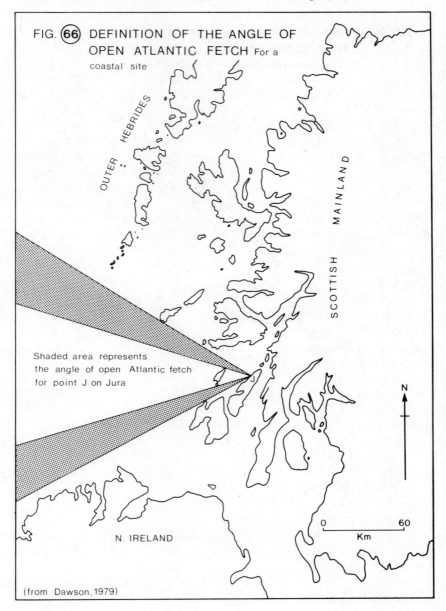

FIG. ⑥⑥ DEFINITION OF THE ANGLE OF OPEN ATLANTIC FETCH For a coastal site

Shaded area represents the angle of open Atlantic fetch for point J on Jura

OUTER HEBRIDES

SCOTTISH MAINLAND

N

0   60
Km

N. IRELAND

(from Dawson, 1979)

Exercise 21: Analysis of indicators of economic development for Latin American countries by application of Spearman's rank correlation coefficient.

Background

   Latin America is the only region of the underdeveloped world that had evolved from colonial status prior to World War II. It is consequently the most advanced of the underdeveloped regions of the world in terms of income *per capita*, death rates and literacy, and is regarded by some as an indicator or harbinger of conditions that will soon prevail in the other underdeveloped regions (Gonzalés, 1967).

Although Latin America has had a relatively long history of development, a wide spectrum of conditions of development is represented in the various countries. An examination of the inter-relationships between certain indicator variables, which are available for each of the countries of Latin America, would be expected to reveal insights into some of the characteristics and problems of economic development. For example, a simple model of economic development suggests that development is accompanied by a fall in the infant mortality rate and in the proportion of the population engaged in agriculture, but by a rise in average incomes and in the gross domestic product (G.D.P.). One would expect, therefore, a positive correlation between the first two variables and between the last two variables, but negative correlations between other combinations of these variables. Correlation coefficients between socio-economic variables may suggest relationships that are important in terms of economic development; alternatively, hypothesized relationships can be tested against the Latin American data.

Eight socio-economic indicators of development are given for twenty Latin American countries in Table 26.

Practical work

1. State, and give a brief explanation of, the direction and strength of relationship that you would expect between the following pairs of variables:
   (i) income *per capita* and calorie intake *per capita*;
   (ii) income *per capita* and percentage of the economically active population in agriculture;
   (iii) manufacturing production index and agricultural production index;
   (iv) agricultural production index and percentage of the economically active population in agriculture.
2. (a) Using Spearman's correlation coefficient, measure the direction and strength of the relationships considered in question 1.
   (b) Test the statistical significance of the relationships.
   (c) Briefly discuss the meaning of each result.
3. Using the appropriate Spearman's correlation coefficients, and assuming those countries with the highest percentage of the economically active population in agriculture are relatively underdeveloped:
   (a) to what extent do the data support the assertion that underdeveloped countries have an undernourished population.
   (b) to what extent do the data suggest that underdevelopment is characterized by high infant mortality rates and low population growth rates?
4. Use Spearman's correlation coefficients to examine the relationships between some other combinations of variables. Give reasons for your selection and a short discussion of your results.
5. Select two relationships from your results (questions 1–4) that have been found to be of no (or low) statistical significance. Consider possible reasons for this result under the headings: (i) the complexities of economic development and (ii) the characteristics of the data.

6. Pearson's correlation coefficient (r) was calculated for the variables involved in question 1 and was found to be:

  (i) variable G with variable $H = r = +0.81$,
  (ii) variable C with variable $G = r = -0.85$,
  (iii) variable D with variable $E = r = +0.44$,
  (iv) variable C with variable $D = r = +0.28$,

Compare the results of Spearman's correlation coefficient (r_s) with the above values for r commenting on any differences and discussing the relative appropriateness of the two coefficients.

TABLE 26. *Indicators of economic development in Latin American countries*

Country	\multicolumn{8}{c}{Variables}
	A	B	C	D	E	F	G	H
Mexico	3.1	67.7	54	183	162	4.9	415	2580
Guatemala	3.2	92.8	68	196	123	4.8	258	1970
El Salvador	3.6	65.5	60	203	152	–	268	2000
Honduras	3.0	47.0	66	162	–	5.4	252	2330
Nicaragua	3.5	53.9	68	226	134	10.8	288	2190
Costa Rica	4.3	77.6	55	119	–	–	362	2520
Panama	3.3	42.9	46	144	–	8.1	371	2370
Cuba	2.0	41.8	42	86	–	9.0	516	2730
Dominican R.	3.6	79.5	56	144	–	–	313	2020
Haiti	2.2	171.6	83	104	–	–	149	1780
Venezuela	3.4	47.9	32	176	175	4.0	645	2330
Colombia	2.2	88.2	54	135	141	5.0	373	2280
Ecuador	3.2	95.6	53	195	–	3.7	223	2100
Peru	3.0	94.8	46	136	154	6.7	269	2060
Bolivia	1.5	86.0*	72	160	–	–	122	2010
Paraguay	2.4	98.0†	54	117	92	3.6	193	2400
Chili	2.4	111.0	28	122	146	5.8	453	2610
Argentina	1.6	60.7	19	122	107	–0.1	799	3220
Uruguay	1.4*	47.4	–	110	98	–	561	3030
Brazil	3.0	170.0‡	58	131	147	4.7	375	2710
Latin America	2.9	–	47	133	129	3.6	421	2570

(From Gonzaléz, 1967.)

Variable A = Population, % annual increment 1958–64 (*1940–50).
Variable B = Infant mortality rate 1960 (*1959; †1945–49; ‡ 1958–62).
Variable C = % economically active population in agriculture *ca.* 1950–60.
Variable D = Agricultural production index 1964 (1958 = 100).
Variable E = Manufacturing production index 1963–64 (1958 = 100).
Variable F = Gross domestic product, growth rate *ca.* 1960–63.
Variable G = Income *per capita* 1961 (U.S. dollars).
Variable H = Calorie intake *per capita* 1959–61.

14

The Form of Relationships and Prediction by Regression

CORRELATION analysis emphasizes the degree to which two sets of data vary together and the direction of the covariation. Correlation does not, however, tell us about the way in which the variables are related, the form of the relationship, or possible anomalies or deviations from the overall form, and it does not enable the prediction or forecasting of values of one variable from a knowledge of the way in which the second variable varies. The technique of *regression* enables consideration of these questions.

Figure 67 describes in graphical form the positive relationship between pH of lake water and distance from a smelter near Sudbury, Ontario, Canada. It has been shown that the strength of this relationship can be quantified by the correlation coefficient, r, as $+0.71$. The generalized form of the relationship is approximated by a straight line, which also highlights any deviations (residuals) from the overall pattern. This *linear relationship* is defined if the intercept (point of intersection of the vertical axis) and the slope of the line are known.

Regression is partly concerned with the construction of 'best-fit' lines describing the form of relationships, such as the one in Fig. 67, and is partly concerned with questions of inference and prediction. The descriptive use of regression is simply *curve-fitting*; the use of regression as a technique of inferential statistics, permitting, for example, the prediction of the pH of lake water from the distance of a lake from the smelter, involves many more assumptions. Linear regression will be emphasized here, although some relationships are obviously *non-linear relationships*. A familiar example of a non-linear relationship is provided by population growth through time, which is often exponential and is approximated by an exponential curve (that is, the rate of increase is constant). Many non-linear relationships can be transformed to linear relationships prior to regression analysis (see Chapter 6 in which the object was to transform a variable to normality, rather than linearity).

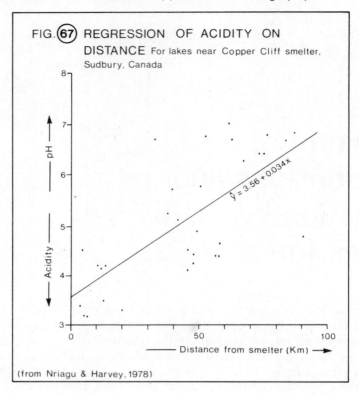

FIG. (67) REGRESSION OF ACIDITY ON DISTANCE For lakes near Copper Cliff smelter, Sudbury, Canada

(from Nriagu & Harvey, 1978)

'Least-squares' regression

In a two-variable situation, if one variable is postulated as being a function of the second variable, then the one variable is said to be dependent on the other. In Fig. 67, pH is termed the *dependent variable* and distance is the *independent variable* (by convention, y (the vertical axis) and x (the horizontal axis) respectively). When it is possible to specify dependent and independent variables, or when one variable is being predicted (y) from another (x), then the appropriate regression line involves the *regression of y on x*. It is particularly important when applying the method of 'least-squares' to regress the dependent variable on the independent variable, because the resulting line is not the same as the line produced by regression of x on y.

Consider Fig. 68A. 'Least-squares' regression fits the curve that minimizes the squares of the distances indicated by short vertical lines. In other words, it minimizes the squares of the y-residuals from the regression line. The regression line itself is described by:

$$\hat{y} = a + bx$$

where $\hat{y}$ = any predicted value of the dependent variable,
$\quad x$ = the corresponding value of the independent variable,
$\quad a$ = the intercept (the value of $\hat{y}$ for which x is zero; the point at which the regression line cuts the y-axis),
$\quad b$ = the slope of the line (the number of units of variable y corresponding to one unit of variable x).

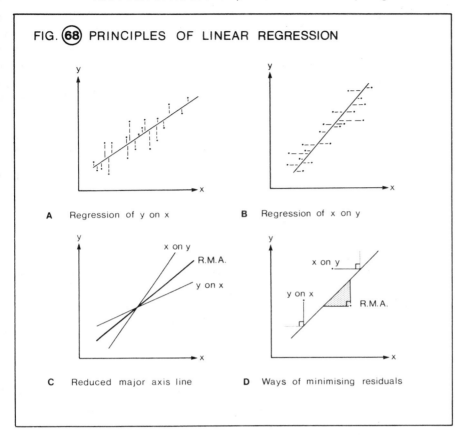

FIG. 68 PRINCIPLES OF LINEAR REGRESSION

A   Regression of y on x

B   Regression of x on y

C   Reduced major axis line

D   Ways of minimising residuals

Thus for any value of x, $\hat{y}$ can be predicted when a and b are known. Because the residuals are minimized in this way, instead of in terms of x-residuals (Fig. 68B), we can be relatively sure that a predicted y value will be close to its true value; that is, the likely error in the prediction is minimized.

If it is required to describe the relationship between two variables that cannot be sensibly regarded as dependent and independent then there is no reason to prefer a regression on y on x to a regression of x on y. In this situation, the line that bisects the angle between the two possible lines (the *reduced major axis line*) is recommended (Fig. 68C). This line minimizes the area of the triangles between individual points and the line (Fig. 68D). All three lines pass through the mean of variable x and the mean of variable y.

The b coefficient (the slope of the line defined by the regression of y on x) is calculated from the formula:

$$b = \frac{n(\Sigma xy) - (\Sigma x)(\Sigma y)}{n(\Sigma x^2) - (\Sigma x)^2}$$

where symbols are the same as those used in the calculation of Pearson's correlation coefficient (r). Given that a least-squares regression line passes through the means of the two variables ($\bar{x}$ and $\bar{y}$), once the b coefficient has been calculated, then the a coefficient can be found by substitution in the basic equation of the regression line ($\hat{y} = a + bx$).

The correlation coefficient can be envisaged as being related to the angle between the lines described by regression of y on x and of x on y. When variability within the data is great (Fig. 69A) then the angle between the regression lines is large; this corresponds to a weak relationship and the correlation coefficient is relatively small (near zero). Figure 69B represents a strong relationship with a correlation coefficient that is relatively large (near one); in this figure the variability is low so that the two regression lines make a small angle.

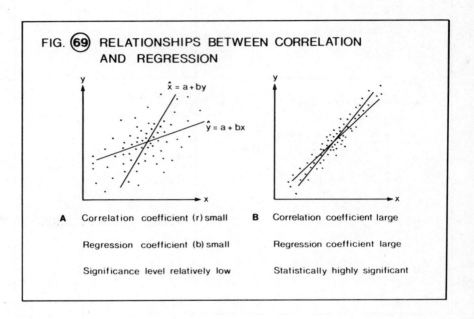

FIG. 69 RELATIONSHIPS BETWEEN CORRELATION AND REGRESSION

$\hat{x} = a + by$

$\hat{y} = a + bx$

A Correlation coefficient (r) small

Regression coefficient (b) small

Significance level relatively low

B Correlation coefficient large

Regression coefficient large

Statistically highly significant

The close relationship between correlation and regression is also shown by the following formula:

$$b = r \cdot \frac{s_y}{s_x}$$

where s_y = the standard deviation of variable y,
s_x = the standard deviation of variable x,
r = Pearson's correlation coefficient.

The standard deviations in the formula are a quantitive expression of the variability factor considered qualitatively and graphically in Fig. 69.

Because of the close relationship between correlation and regression, the statistical significance of the slope of a regression line (that is, whether the slope is sufficiently different from zero) is tested in the same way as the significance of the correlation coefficient (that is, whether r is sufficiently different from zero).

Calculations

The regression line in Fig. 67 was calculated as follows, with pH as the dependent variable (y) and distance as the independent variable (x), using the data given in Chapter

3. The slope of the regression line is:

$$b = \frac{n(\Sigma xy) - (\Sigma x)(\Sigma y)}{n(\Sigma x^2) - (\Sigma x)^2}$$

where $\Sigma y = 162.26$

$\Sigma x = 1427.1$

$\Sigma xy = 7957.853$

$(\Sigma x)^2 = 2,036,614.4$

$(\Sigma x^2) = 84,895.83$

that is

$$b = \frac{32\,(7957.853) - (1427.1)(162.26)}{32\,(84,895.83) - (2,036,614.4)}$$

$$= \frac{254,651.29 - 231,561.24}{2,716,666.5 - 2,036,614.4}$$

$$= \frac{23,090.05}{680,052.1} = 0.033\,95$$

The coefficient a is next calculated by substitution of three known quantities in the formula:

$$\hat{y} = a + bx$$

where $b = 0.033\,95$

$$\bar{x} = \frac{\Sigma x}{n} = 44.5969$$

$$\bar{y} = \frac{\Sigma y}{n} = 5.0706$$

Thus $a = \bar{y} - b\bar{x}$

$$= 5.0706 - (0.033\,95)(44.5969)$$

$$= 5.0706 - 1.5140 = 3.5566$$

The equation for the regression line is therefore $\hat{y} = 3.556\,73 + 0.033\,95x$, from which other values of y can be predicted. Although a regression line (such as that in Fig. 67) can be drawn once the intercept and the values for $\bar{x}$ and $\bar{y}$ are known, it is advisable to calculate a third point (which, if the calculations have been carried out correctly, should lie in a straight line). In this instance, the predicted acidity ($\hat{y}$) corresponding to a distance of 100 km is found from:

$$\hat{y} = a + bx = 3.556\,73 + (0.033\,95)(100.0)$$

$$= 6.9517$$

It can be seen from Fig. 67 that the calculation is correct and that we have in fact predicted the acidity of a lake found at 100 km distance from the smelter. It should be pointed out, however, that it is dangerous to predict values from regression lines that have been extrapolated from beyond the area of the diagram for which there is good control by data points. It must also be emphasized that predictions only apply to lakes around this particular smelter, that is to the population of lakes from which the measured lakes were sampled.

The slope of the regression line is statistically significant as it has been shown that the correlation coefficient is significantly different from zero. We conclude, therefore, that the slope of the regression line in Fig. 67 is greater than is likely to have resulted by chance or to have been found from sampling from lakes that exhibit 'no relationship' between pH and distance from the smelter.

An example of how regression may be used further in a predictive way is provided in Figs. 70 and 71. Figure 70 shows the relationship between tree growth index (derived from the width of annual rings) and summer temperature index (derived from June and July temperatures) for Scots pine *(Pinus sylvestris)* near the tree-line in southern Norway. Each data point represents a particular year from 1901 to 1950, the period over which meteorological data are available for this area. The regression equation is given and Pearson's correlation coefficient is $r = +0.77$. The strength of this correlation coefficient can be attributed to summer temperature being the limiting environmental factor near the tree-line at high altitudes and high latitudes, where moisture is normally adequate for tree growth. Using the regression line in Fig. 70, summer temperatures can be predicted from tree growth. Assuming that the same controls on Scots pine existed in the past it is

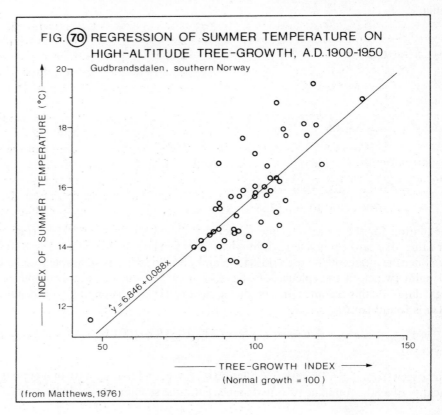

FIG. (70) REGRESSION OF SUMMER TEMPERATURE ON HIGH-ALTITUDE TREE-GROWTH, A.D. 1900-1950

Gudbrandsdalen, southern Norway

$\hat{y} = 6.846 + 0.088x$

INDEX OF SUMMER TEMPERATURE (°C)

TREE-GROWTH INDEX
(Normal growth = 100)

(from Matthews, 1976)

possible to predict summer temperatures back to the beginning of the eighteenth century using values of tree-growth from the long tree-growth series in Fig. 71 as the independent variable. In this way, information on climate can be obtained for periods when few instrumental records are available (Matthews, 1976).

The final example indicates how an analysis of residuals may form part of a regression analysis. Figure 72A summarizes the form of the relationship between rural farm population density and mean annual precipitation in Nebraska (discussed in the context of correlation in Chapter 13). The residuals (deviations of data points from the regression line in Fig. 72A) are mapped in Fig. 72B. The map indicates those areas of Nebraska where the relationship is least reliable or where predictions of rural farm population density are most likely to be in error. It therefore points to those areas where additional factors to precipitation are necessary to explain the level of rural farm population density. In the centre of the state, rural farm population densities tend to be overpredicted (that is, residuals are negative and the predicted values are higher than were observed in reality); towards the east and west there is a tendency for underprediction (where residuals are positive and the predicted values are lower than the observed values). It should be noted, however, that there is one extreme residual in the west of the state, perhaps the result of a unique factor.

The coefficient of determination (r^2)

This coefficient, simply the square of the correlation coefficient, is a useful aid in the interpretation of any regression analysis. It measures the proportion of the variability in one variable that can be accounted for, determined from, predicted or 'explained' by variability in the second variable. In other words, it is a measure of the 'goodness-of-fit' of a regression line, $r^2 = 1.0$ denoting a perfect fit and the possibility of predicting with certainty.

The regression in Fig. 67 has a coefficient of determination of $r^2 = 0.51$, which indicates that distance from the smelter accounts for just over 50 % of the variability in the acidity of the lakes. This in turn indicates that other variables, such as the volume of the lakes, account for almost as much of the variability in acidity. The regression in Fig. 70 has a coefficient of determination of 0.59, which indicates that some 40 % of the variability in tree growth is accounted for by variables other than summer temperature. In both examples, therefore, there remains considerable uncertainty in any predictions. This uncertainty could be measured precisely using confidence intervals around the predicted values of the dependent variable, a topic covered in some textbooks and sometimes represented as a confidence band around regression lines.

Limitations of linear regression

Use of regression for inferential statistical purposes is limited by its many assumptions. Some of these are:
1. Interval-scale data are required.
2. The relationship is assumed to be linear, or the data must be transformed to linearity prior to analysis.

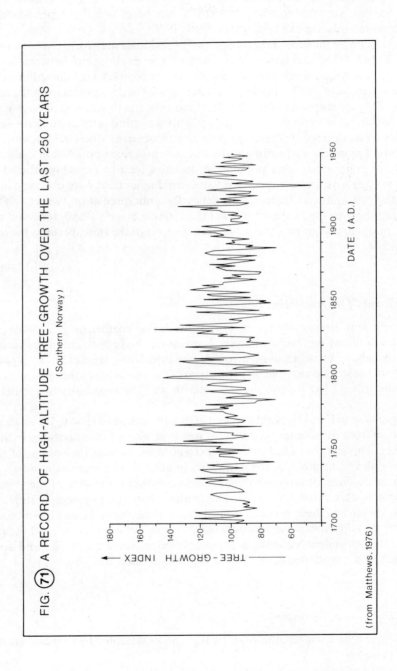

FIG. ⑦1 A RECORD OF HIGH-ALTITUDE TREE-GROWTH OVER THE LAST 250 YEARS

(Southern Norway)

TREE-GROWTH INDEX

DATE (A.D.)

(from Matthews, 1976)

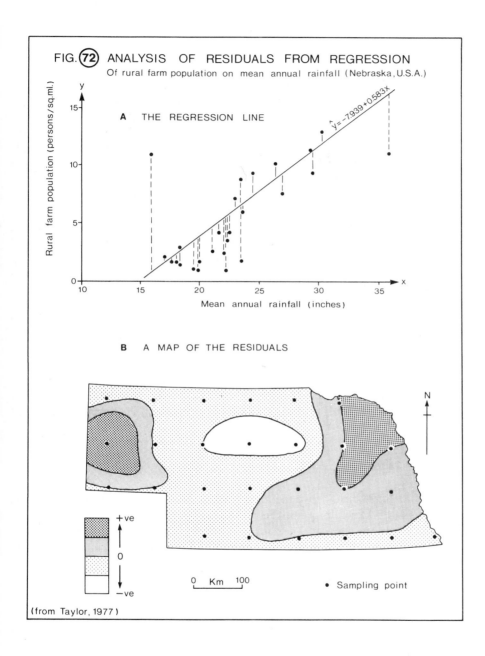

FIG. (72) ANALYSIS OF RESIDUALS FROM REGRESSION

Of rural farm population on mean annual rainfall (Nebraska, U.S.A.)

A THE REGRESSION LINE

$\hat{y} = -7.939 + 0.583x$

Rural farm population (persons/sq.ml.)

Mean annual rainfall (inches)

B A MAP OF THE RESIDUALS

N

+ve

0

−ve

0 Km 100

• Sampling point

(from Taylor, 1977)

3. Homoscedasticity is assumed. That is, the population distributions of y for every value of x are assumed to possess equal variability.
4. Any errors in the measurement of the x variable are assumed to be small in relation to any errors in the measurement of variable y.
5. Independent random sampling is assumed.

Exercise 22: Prediction of annual temperatures from sea-ice conditions off Iceland using regression analysis.

Background

Historical evidence of the extent of sea-ice off the coast of Iceland has survived from the time of settlement by Norsemen in the ninth century. Although chronicle writing began several centuries after settlement, the historical accounts provide valuable evidence of sea-ice conditions, and hence climate, since that time (with the notable exception of the fifteenth century). From the beginning of the seventeenth century, records are sufficiently detailed to enable the reconstruction of a continuous record of ice incidence in months per year (Fig. 73A); from earlier times, only a generalized picture of ice conditions is available (Fig. 73B).

Bergthórsson (1969) has shown how regression analysis can be used to estimate, from the incidence of sea-ice, substantial changes in the mean annual temperatures in Iceland over the last 1000 years – the whole period of human settlement on the island and a record unequalled anywhere in the world. The first stage of the analysis involves the relationship between the incidence of sea-ice and mean annual temperature since 1846, when meteorological data are available from local meteorological stations. The strength of the relationship is estimated by correlation, and linear regression is used to describe the overall form of the relationship. The second stage of the analysis uses the regression line and regression equation to predict the past history of temperatures in Iceland from the long record of sea-ice conditions.

It transpires that a knowledge of the mean annual temperatures derived in this way helps to explain some of the features of human history in Iceland and in neighbouring areas, such as Greenland where settlements were established at the end of the tenth century but did not survive the 'Little Ice Age'. The Greenland settlers appear to have died out by the beginning of the sixteenth century.

Practical work

The tabulated data (Table 27) give the mean annual temperature for Iceland based on weather stations at Stykkishólmur and Teigarhorn, and the incidence of sea-ice off the coast of Iceland in months per year. The data are from the years 1846–1919; years in which no sea-ice was observed have been omitted, and the data are not listed in their order of occurrence year by year.

    1. (a) Which of the two variables is the dependent variable in the context of the exercise?

       (b) Draw a scatter graph of the relationship between the two variables.

       (c) Measure the strength of the relationship using Pearson's correlation coefficient.

       (d) Calculate the regression equation appropriate for the prediction of mean annual temperatures from sea-ice conditions.

       (e) Draw the best-fit line on your graph.

       (f) Briefly describe and explain the direction, strength and form of the relationship.

    2. (a) Is the slope of the regression line statistically significant?

       (b) Calculate and assess the coefficient of determination for this relationship.

    3. The following questions should be answered in two ways: firstly, using the regression

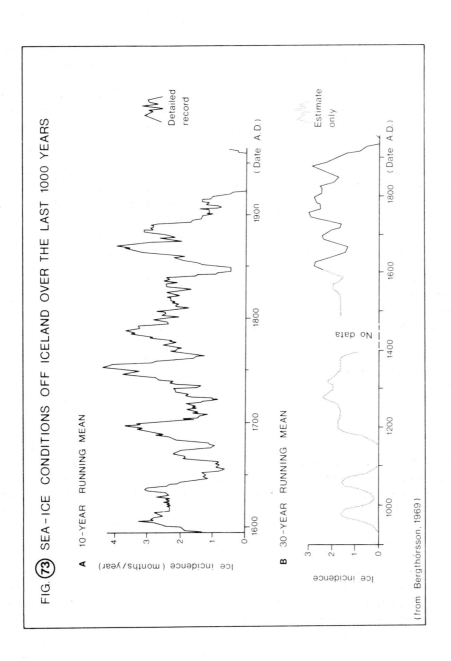

FIG. (73) SEA-ICE CONDITIONS OFF ICELAND OVER THE LAST 1000 YEARS

(from Bergthorsson, 1969)

TABLE 27. *Temperatures and sea-ice conditions off Iceland in recent times*

Mean annual temperature (°C)	Sea-ice incidence (months/year)	Mean annual temperature (°C)	Sea-ice incidence (months/year)
4.4	0.5	3.7	2.8
4.1	0.8	3.5	2.8
4.2	0.9	3.4	2.9
4.1	1.1	3.2	2.8
3.9	0.6	3.2	2.2
3.9	0.9	3.1	2.4
3.8	0.4	2.4	2.9
3.5	1.0	2.6	2.9
3.2	0.7	3.3	3.1
3.1	0.9	3.6	3.2
3.1	0.7	3.6	3.3
3.0	0.8	2.6	3.3
2.8	0.6	2.2	3.5
2.7	0.2	1.0	3.2
2.6	0.6	1.8	4.6
2.4	0.9	2.1	4.6
2.2	1.4	2.0	4.0
2.4	1.2	2.2	4.8
2.6	1.2	2.5	4.1
2.5	1.3	2.7	4.6
2.9	1.3	2.9	5.0
2.8	1.8	3.8	4.6
3.0	1.7	2.7	5.1
3.2	1.2	1.2	5.5
3.3	1.5	1.2	6.0
3.4	1.3	1.8	6.2
3.4	1.6	2.4	7.2
3.4	1.8	1.0	7.3
3.4	1.9		

(From Bergthórsson, 1969)

line that you have drawn on your graph, and secondly, directly from the regression equation that you have calculated.

(a) If sea-ice survives for 6 months, what mean annual temperature is predicted?

(b) What mean annual temperature is indicated if no sea-ice is observed in any year?

(c) What change in mean annual temperature would be expected if, over a period of years, the incidence of sea-ice increased by one month?

4. In what ways are the following statements incorrect?

(a) The regression analysis enables one to say that in years with 3 months of sea-ice there will be a mean annual temperature of 2.78°C.

(b) The fitted regression line can be used to predict the incidence of sea-ice in years when mean annual temperatures are 3.0°C.

(c) The regression equation should be used in preference to the regression line when predicting mean annual temperatures from sea-ice incidences in excess of 10 months.

5. Using the results of your regression analysis, in conjunction with Fig. 73, answer the following questions as far as you are able.

(a) What was the lowest mean annual temperature experienced in Iceland over the last 1000 years?

(b) What has been the degree of change in mean annual temperatures in Iceland during the twentieth century?

(c) What have been the main changes in the mean annual temperature of Iceland since A.D. 900?

6. Comment on the possible limitations of this type of analysis.

7. To what extent can the evidence from this analysis be used in support of the contention that the settlement of Iceland and Greenland and the subsequent extinction of the Greenland population were caused by changes in summer temperature conditions.

Exercise 23: Application of regression to the description and analysis of urban population densities in London and Chicago.

Background

One of the early contributions to quantitative Geography established a relationship between population density and distance from a city centre (Clark, 1951). Clark showed that the relationship has a similar form in twenty cities including Berlin, Boston, Budapest, Dublin, Manchester, Melbourne, Oslo, Philadelphia and Vienna; since Clark's study many more cities have been shown to exhibit similar negative exponential relationships (Berry *et al.*, 1963). This form of relationship is described by the regression equation:

$$\hat{\log_e} y = \log_e a - bx$$

where $\hat{\log_e} y$ = the natural logarithm of population density (predicted value of the dependent variable),

$\log_e a$ = the natural logarithm of population density in the city centre (the intercept of the regression line),

b = the slope of the regression line (population density gradient),

x = distance from the city centre (independent variable).

In other words, there is a constant *rate* of decline in population density with distance from the city centre, which is described by a linear relationship between the natural logarithm of population density (the dependent variable) and distance from the city centre (in-dependent variable). A graphical representation of this form of relationship is shown for Los Angeles, California, U.S.A., in Fig. 74.

According to Berry *et al.* (1963) the negative exponential form is a logical outcome of urban land-use theory, but the intercept (city centre density) and the slope (density gradient) vary from city to city. In this exercise the aim is to compare the change in the intercept and slope of the relationship during the growth of two 'Western cities'—London and Chicago—and to relate these changes to some alternative models of city growth. Figure 75 A to C is a diagrammatic representation of three possible patterns of change in the density–distance relationship through time. These are:

(a) the 'uniform growth' model, in which growth in population density occurs equally at all distance from the city centre;

(b) the 'suburban growth' model, in which peripheral growth exceeds city-centre growth;

FIG. ⑭ POPULATION DENSITIES IN LOS ANGELES, U.S.A. (1940)

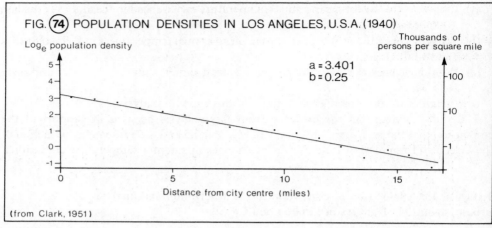

(from Clark, 1951)

FIG. ⑮ MODELS OF CHANGE IN URBAN POPULATION DENSITY

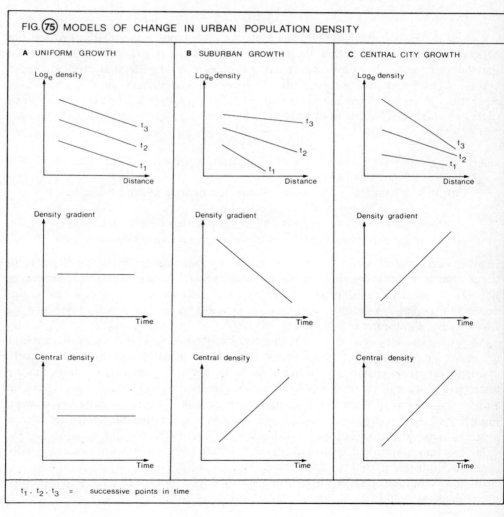

t_1, t_2, t_3 = successive points in time

(c) the 'city-centre growth' model, in which the central density increases faster than densities at the periphery.

It should be recognized that the three models are distinguishable in terms of the intercept and slope of regression lines.

Practical work

1. Examine Fig. 75 and, in a similar format, construct a fourth model of changes in the density–distance relationship through time.

2. (a) Using the data given in Table 28, draw up accurate graphs of the relationship between population density and distance from the city centre for:
   (i) London (1801–1939),
   (ii) Chicago (1860–1950).

TABLE 28. *Results of regression analysis of population densities in Chicago and London*

Year	Regression coefficients	
	Central density ('000's per mile^2)	Density gradient
London (after Clark, 1951)		
1801	290	1.35
1841	800	1.40
1871	290	0.65
1901	210	0.45
1921	180	0.35
1939	80	0.20
Chicago (after Berry et al., 1963)		
1860	30	0.91
1880	97	0.79
1900	100	0.40
1920	73	0.25
1940	71	0.20
1950	64	0.18

The graphs should be drawn on semi-logarithmic graph paper (e.g. Fig. 74) and the regression equation should be used in the construction of the regression lines.

(b) Describe the graphs.

3. Evaluate the models in Fig. 75 in relation to the evidence from London and Chicago.

4. The aim of this question is to test further the extent to which one of the models in Fig. 75 gives a true picture of the growth of London and Chicago. Use whichever of the models appears to be the best representation of reality.

(a) Draw graphs of density gradient against time and of central density against time for London and Chicago.

(b) Note that the models in Fig. 75 suggest linear relationships in each case. Test, using regression analysis, whether the slope of each regresssion line is significantly different from zero, and hence whether the model is a good one.

5. Construct an improved, revised model of the density–distance relationship through time (in the format of Fig. 75) in the light of your answers to questions 1 to 4.

6. It is generally held that changes in the population density in London and Chicago owe much to the provision of mass transport systems, such as steam trains, underground (in the case of London) and surface electric trains (in the case of Chicago).

(a) Discuss how the efficiency and cost of mass transport may influence changes in population density with distance from the city centres.

(b) Given that many non-Western cities have not had the benefit of efficient mass transport systems, discuss which model is most likely to 'fit' non-Western cities.

7. The highest city centre density ever recorded was estimated at greater than 350,000 per square mile from Lower East Side, New York, in 1900, and archaeologists estimate that from the earliest times cities have supported maximum densities of less than half this value (Clark, 1951). In the light of the central densities predicted by the regression lines for London (particularly the value for 1841) what is the implication for this kind of analysis?

8. If the data were expressed in '000's per km^2, rather than square miles, would the following remain the same:

(a) the value of a (the intercept of the regression line);

(b) the value of b (the slope of the regression line);

(c) the value of r (the correlation coefficient);

(d) the statistical significance of the slope of the regression line.

15

Conclusion: Choice and Limitations of Statistical Techniques

"For the most part, Statistics is a method of investigation that is used when other methods are of no avail; it is often a last resort and a forlorn hope. A statistical analysis, properly conducted, is a delicate dissection of uncertainties, a surgery of suppositions."   M. J. Moroney (1956), p.3.

Choice of a statistical technique for hypothesis testing

There is no real substitute for practical experience in choosing an appropriate technique to assist in solving a particular problem. A simple scheme to help in this choice is given in Table 29, which includes those techniques that are covered in this manual and are suitable for statistical hypothesis testing; other useful techniques can be found in the textbooks listed at the end of the manual under 'Further Reading'. The table employs four criteria to determine which technique should be used:
  (a) the nature of the hypothesis;
  (b) the level of measurement of the data;
  (c) the number of samples;
  (d) the nature of the samples (dependent or independent).
First priority must be given to the nature of the hypothesis, which is determined by the purpose of the investigation. Some hypotheses cannot be tackled by the use of techniques given in the manual. Examples have been given of hypotheses in which we are interested but which cannot be tested directly, such as hypotheses about cause and effect relationships discussed in the context of correlation. However, there are few problems involving data where the judicious use of statistics does not clarify some issue in connection with the problem. Thus although a high correlation coefficient does not indicate a cause, it does indicate that a causal relationship is possible and may prompt a search for causes.

Table 29 includes three types of hypothesis that can be tested *directly* using the techniques in the manual. These are:
  (a) hypotheses about differences;
  (b) hypotheses about the strength of relationships;
  (c) hypotheses about the form of relationships.
In order for a hypothesis to be tested statistically it must be phrased in a testable form;

TABLE 29. Choice of an appropriate technique

Level of measurement requirement	Questions about differences					Questions about the strength of relationships		Questions about the form of relationships
	1 sample	2 independent samples	2 dependent samples	> 2 independent samples	> 2 dependent samples	2 samples	> 2 samples	(2 samples)
Nominal scale	One-sample χ^2 test	Two-sample χ^2 test	—	χ^2 test for k samples	—	Contingency coefficient (C)	Contingency coefficient (C)	—
Ordinal scale	—	Mann–Whitney test (U) Kolmogorov–Smirnov test (D)	Wilcoxon matched-pairs signed-ranks test (T)	Kruskal–Wallis analysis of variance by ranks (H)	Friedman two-way analysis of variance (χ^2)	Spearman's rank correlation coefficient (r_s)	—	—
Interval scale	One-sample Student's t-test	Student's t-test for two independent samples	Student's t-test for two dependent samples	—	—	Pearson's correlation coefficient (r)	—	Regression

that is, in the form of a null hypothesis (or a hypothesis of 'no difference'). Reasons for this requirement were given at the end of Chapter 9. The danger here is that the hypothesis actually tested may not be the one that is required for the purpose of the investigation, unless great care is taken over the formulation of hypotheses. When a decision is made regarding the null hypothesis, further care is required in the correct interpretation of the result. Recall, for example, that 'failing to reject' a null hypothesis does not mean the same thing as 'accepting' the null hypothesis (see Chapter 9).

The level of measurement of the data is important, because particular tests require data at particular levels of measurement. Three levels of measurement are recognized in the table; these are, in order of increasing precision of measurement:

    (a) nominal scale measurement;
    (b) ordinal scale measurement;
    (c) interval scale measurement.

Although it is not possible to use data based on a low-precision measurement scale (e.g. nominal scale measurements) in a statistical test demanding a high-precision level of measurement (e.g. Student's t-tests), it is possible to employ high-precision data (e.g. interval scale measurements) in a statistical test with less demanding requirements (e.g. χ^2 tests) if the data are appropriately 'degraded' to the lower level. It should be borne in mind, however, that information is lost if the latter practice is adopted and that, as a general rule, a test should be carried out at the highest level of measurement that is available (provided that other assumptions of the test are met).

Different tests are often necessary for different numbers of samples. Table 29 distinguishes between:

    (a) one-sample tests;
    (b) two-sample tests;
    (c) tests appropriate for more than two samples.

Some tests, such as the extremely flexible χ^2 tests, can be used, with only slight modifications, for any number of samples, while others, such as the Mann – Whitney test, are strictly for a specified number of samples. Although a modified Kolmogorov – Smirnov test is available for one-sample testing, only the two-sample test has been introduced in the manual. An elementary mistake, to be avoided, is the confusion of the number of samples with the sample size. An adequate sample size is, of course, required for all statistical tests, and may be an additional criterion for selection of a particular technique.

Lastly, the table recognizes the need for different tests for:

    (a) independent samples;
    (b) dependent (matched) samples.

At first sight, it may be difficult to decide whether or not two samples are independent, but it is important to be certain of the nature of the sample for two reasons. Not only is it incorrect to apply one category of test to the wrong kind of sample, but failure to recognize dependence will result in the loss of important, controlled information (see Chapters 9 and 12). Although tests that are suited to dependent samples are based on matched-pairs or matched-sets of data, they still require the pairs or sets to be independent of each other.

Additional assumptions of individual tests have been outlined in their respective chapters and the more stringent assumptions of the parametric tests (those along the bottom row of Table 29) are discussed further in the following section.

On the limitations of a statistical approach

Some limitations of the various statistical techniques that have been considered in the present and preceding chapters will be already apparent. The purpose of this section is to summarize and discuss these limitations. An understanding of limitations is essential for evaluating the use of any technique to be found in the geographical literature, and is particularly necessary if a technique is to be used in a dissertation or in any other project.

Perhaps the greatest limitation of statistics is that they are *techniques* and as such are applied for particular purposes in the context of particular problems; they are one set of tools for use in the pursuit of knowledge by scientific method. Two variations on this theme require emphasis here. First, elaborate techniques are no substitute for precise and accurate data, a point that can be summed up by stating that statistical analyses are only as good as the data on which they are based. Second, statistical significance must not be confused with geographical significance. For example, Table J shows that if a sample size of 100 is employed, then a correlation coefficient of $r_s = 0.2$ is statistically significant at the 5 % significance level. But a correlation coefficient of 0.2 indicates that only 4 % of the variability of one variable is accounted for by variability in the second variable (see the coefficient of determination in Chapter 14), which in turn indicates that this level of correlation provides us with a very low level of explanation in real terms. Thus one should not expect statistics to 'produce' new Geography, only to act as an aid in the imaginative analysis of carefully collected data. In short they are a means to an end, not an end in themselves.

Probability is the central theme of inferential statistics. Throughout this manual a 5 % significance level has usually been employed in the testing of hypotheses, and a 5 % significance level is commonly used in geographical applications generally. The precise meanings of the term 'significance level', and related terms such as 'confidence level', have been emphasized in the manual. However, the *arbitrariness* of a chosen significance level is a very great limitation of statistics, for this makes what appears at first sight to be an objective procedure dependent on judgement. For example, there is no definite level at which a difference between two samples establishes with certainty a difference between the corresponding populations; there is always a possibility of a wrong decision, albeit at a known level of probability. On the other hand, some geographers have argued that the commonly used significance levels are too stringent, that relationships and differences that are significant at the 5 % level are obvious anyway (with the consequence that statistics are not required to tell us so) and that many weak relationships that would be thrown out using a 5 % significance level should be retained for deeper investigation (Gould, 1970). The arbitrariness of the significance level also means that inferential tests are easily abused by choice of significance level *after* the test has been performed. The last point is more a human limitation than a limitation of statistics, however. A significance level should be chosen *before* a test is carried out (with due regard to the seriousness of making a wrong decision) or the precise level of significance should be quoted for each test. If the latter procedure is adopted then the arbitrariness remains to be resolved by the reader rather than the investigator.

All the techniques considered in the manual are limited by one or more *assumptions* relating to the sampling scheme, the data and/or the underlying population. Random sampling has been assumed for most of the techniques, although other objective and unbiased sampling schemes (such as systematic sampling) are usually considered equally valid bases for statistical inference. A related point, which has caused considerable concern to quantitative geographers in recent years, is the difficulty of distinguishing between population and sample in some applications. For example, in Exercise 17 on the planning regions of Mid-Glamorgan, South Wales, socio-economic data for the 113 wards were used to test whether or not the four regions differed. Some would argue that we used the population for each region (rather than samples). If this was so, then inferential statistics (which make inferences about underlying populations) would be inappropriate. The problem can be circumvented by considering a *theoretical population*, from which the actual Mid-Glamorgan situation is considered to be one possible sample outcome. When we asked 'are the differences between the regions greater than are likely to have resulted by chance?' we were in effect saying that there are many possible outcomes that could occur by chance (in theory) and testing whether or not the measured situation might be one of them. In other words, is the observed pattern a non-random outcome or not, or can the observed pattern be regarded as the result of a random (stochastic) process or not? This way of looking at the problem has been termed inferential statistics in a *natural sampling context* (Silk, 1979) to distinguish it from the use of inferential statistics in an *artificial sampling context*; an example of the latter being the random sampling of a number of individuals from a real-world population of individuals.

A major distinction has been made between *parametric and non-parametric* statistics because there are major differences in their statistical assumptions. In particular, the more demanding requirements of the parametric tests (such as normality, linearity and homoscedasticity) present a problem. When all of these assumptions are met by the data under analysis, a parametric test is preferable to a non-parametric alternative. When the assumptions are not met, then four alternatives are available:

(a) apply the parametric test without satisfying its assumptions;
(b) apply the parametric test after data transformation(s);
(c) apply a non-parametric test;
(d) apply both parametric and non-parametric tests in parallel.

There is evidence to suggest that many parametric tests are quite robust, meaning that they are insensitive to moderate violations of their assumptions (Norcliffe, 1977). More research is necessary, however, on the degree of departure from the ideal which can be tolerated. There are as yet no well-founded guidelines for the user to follow. Transformations may sometimes satisfy the normality and linearity assumptions of parametric tests, but only fairly simple transformations (such as the use of logarithms and square roots) are usually interpretable in a meaningful way. The non-parametric tests have many attractive features that recommend them for geographical application, but less is known about their limitations, which have not been so thoroughly explored as the limitations of the parametric tests. The most important advantage of the non-parametric techniques is their ready application to data that are only available at low levels of measurement (such as nominal and ordinal scales). Their most important disadvantage lies in their use of less information about the individuals being analysed; when all the assumptions of the equivalent parametric test are met, the power-efficiency of the equivalent non-parametric test is therefore less (that is, a larger sample size is necessary in the case of a non-parametric test to be equally effective in the rejection of null hypotheses). One method of overcoming such difficulties is to apply both parametric and non-parametric tests to the same data. The difficulty of using both kinds of test in parallel arises if they disagree, a result that may well arise in borderline cases where a clear-cut decision would be most valuable.

The *multivariate* nature of the real world places further constraints on the use of statistics. This is clearly demonstrated with reference to the techniques of correlation and regression. There are often many more than two variables interacting in any real-world situation so that a consideration of only two variables contributes, at best, only a partial explanation. Although the manual contains only three techniques that are applicable to more than two samples (χ^2 tests, the Friedman test and the Kruskal – Wallis test), there is a vast field of statistics – multivariate statistics – which takes this kind of analysis much further. Nevertheless, many insights can be obtained into complex real world problems by the careful use of one-and two-sample tests.

Most of the statistical techniques in the manual are appropriate for independent samples, although some tests for two or more dependent samples have been considered and other issues of non-independence were introduced in the context of time-series analysis in Chapter 5. Time-dependence is related in principle to *spatial-dependence* (spatial autocorrelation) which is manifest as gradients and clusters on maps, and lies at the very roots of Geography. In Geography, the central concern is often with spatial populations distributed over the earth's surface, whereas statistical theory is based largely on abstract statistical populations distributed along a measurement scale. Thus most statistics can be said to ignore spatial co-ordinates. Can techniques designed for non-spatial populations be transferred to spatial populations? A random sample from a map is not a random sample from a non-spatial sampling frame unless the individuals are randomly distributed in space. Yet the phenomenon of spatial autocorrelation describes the common situation of non-random spatial distributions. Given the existence of non-random spatial distributions, the *scale* at which a problem is considered and the choice of *areal units* will lead to different results with the same technique. These fundamental points have often been neglected in the enthusiasms of the application of statistics to Geography and have led some geographers to stress the need for *spatial statistics* (rather than the application of statistical techniques in geographical contexts). In this respect, the design and investigation of measures of spatial autocorrelation form one of the most important frontiers of quantitative Geography today.

The above discussion shows that the limitations of a statistical approach to Geography are not inconsiderable; they are the limitations of particular statistical techniques, of statistics in general and of statistics in Geography. However, wherever anything is measured and wherever there is an attempt to assess variability in the form of numbers, there is the necessity to define our uncertainty and to specify significance. To obtain such benefits while avoiding many pitfalls requires some effort. The encouragement of clear logical thinking in the application of statistics, such as in the formulation and testing of hypotheses, is an additional bonus. At the present time the debate in Geography is not *whether* statistics should be applied but *how* they should best be used in the accumulation of reliable geographical knowledge. What can be said with certainty is that the only way to proceed is for more geographers to become competent in statistics (and vice versa).

Exercise 24: Choosing an appropriate technique for particular purposes.

Background

This exercise is concerned with choosing techniques, rather than carrying them out. There may be a number of alternative techniques that are equally suited to the solution of a particular problem posed. More usually, however, one particular technique is more suitable than any other included in the manual. In answering the questions the most important point is to give full justification for your choice of technique, to give reasons for rejection of other possibilities, and to point out any limitations or uncertainties remaining after the chosen technique has been applied.

Practical work

1. (a) A research worker in Ghana is studying the distribution of an eye infection known as 'river blindness'. He knows the number of infected people in a random sample of 100 people living on a river flood plain and in a second random sample of 200 people living on the neighbouring plateau. How can he test the hypothesis that the incidence of infection on the flood plain is the same as on the plateau?

(b) In a second survey, the same worker has collected data on the level of infection in a sample of twenty villages. He knows the proportion of the population infected in each village and the distance of each village from the river. (i) How could he test whether or not there is a significant relationship between the level of infection and distance from the river? (ii) How could he best estimate the likely level of infection at a village located at a known distance from the river?

2. (a) A sample survey of chalk grassland was conducted in a single valley in south-east England, based on 190 quadrats located at the intersection of a grid. A previously conducted, detailed, regional survey revealed that there are, on average, fifteen species per square metre in chalk grassland generally. How can a decision be reached on whether or not the sample valley is suitable for: (i) a detailed case study of the chalk grassland ecosystem; and (ii) the establishment of a nature reserve?

(b) The same sample survey showed that two particular plant species were found growing together in fifty quadrats; in ninety quadrats neither of the species were found; in twenty-five quadrats the first species occurred, but not the second species. What statistical test could the investigator carry out to help him decide whether or not the two species belong to the same plant community?

3. A recent earthquake in Turkey resulted in the following number of deaths in fifteen villages located at increasing distances from the earthquake epicentre: 520, 410, 320, 310, 50, 210, 250, 400, 100, 100, 20, 50, 80, 90, 200.

(a) If no further information is available, how can these data be used to decide if proximity to the earthquake epicentre increased significantly the number of casualities?

(b) In what ways could the answer be improved if the following information is known: (i) the population of the villages; (ii) the precise distance of each village from the epicentre; (iii) a map showing the location of the villages?

4. A random sample of thirty-three urban counties and nineteen rural counties in the U.S.A. showed that at a presidential election the mean percentage voting Democrat was 57% ($s = 11\%$) and 52% ($s = 14\%$), respectively.

(a) How might one test the contention that the level of Democratic support was higher in urban counties than in rural counties?

(b) Given that in the previous election the urban counties of the U.S.A. voted, on average, 65% Republican and 34% Democrat, how could the sample data be used to decide whether or not there was a significant swing to the Democrats in urban areas?

5. Religious affiliation and voting preference were inter-related by means of a sample survey in an English city. The numbers of people involved in each category were:

Voting preference	Protestants	Catholics	Others
Conservatives	126	61	38
Labour	71	93	69
Others	19	14	27

(a) On the basis of these findings, how could the idea that religious affiliation influences voting behaviour be tested?

(b) How could these data be used to determine whether Catholics vote differently from the remainder of the population?

6. In a study of rates of weathering, the thickness of weathering rinds (measured to the nearest 0.01 mm) were examined on boulders deposited by a glacier in the mountains of British Columbia. Fifty boulders were measured of each of four rock types on each of four moraines of known age. How could each of the following hypotheses be tested:

(a) Rock type influences weathering rind thickness, on surfaces of the same age?

(b) Weathering proceeds at a constant rate, for a particular rock type?

(c) Weathering rate does not differ between rock types?

7. Five alternative sites are being considered for a new airport close to a major city in the United Kingdom. A random sample of 500 citizens are asked to place the five sites in order of preference.

(a) What statistical test could be used to determine whether or not there is agreement amongst the public as to their preference?

(b) What test would be appropriate if there were only two prospective sites?

8. A tidal study is being made in a shallow bay in the West Indies, using coloured pebbles and observing their direction of movement. Observations were made on the pebbles over two periods, each of 60 consecutive days. The number of pebbles that had moved in each period were:

Direction of movement	Period 1	Period 2
N	9	10
NE	15	13
E	8	9
SE	6	6
S	6	5
SW	2	3
N	4	6
NW	10	8

(a) How can a test be made of whether or not there was a significant movement in any one direction, in either period?

(b) What test is appropriate to decide whether or not there was a significant difference between the results in the two periods?

9. Four districts (A to D) in Nigeria have approximately the same area and population but different numbers of primary schools (32, 30, 48 and 50, respectively). The chiefs in districts A and B feel that their districts are neglected by central government, whereas in

districts C and D the chiefs argue that the differences in the selection processes are 'accidental' and are not due to government bias (although the prime minister was born in region D). How could the arguments be analysed statistically?

10. (a) The organic matter content of the surface horizon of soil in an area of heathland was sampled at fifty sites and a comparable sample was taken from a neighbouring area of pine forest. Organic matter content was expressed in grammes of carbon per 1 kg of soil (wet weight). What statistical test could be used to determine whether or not the organic content of the heathland soil differs from that of the pine forest?

(b) A second study was made of the organic matter changes that accompanied the afforestation of an area of heathland by a pine plantation. Prior to afforestation, fifty sites were selected and from each site a soil sample was retained. Twenty-five years after afforestation, soil samples were taken from the same sites. How could a statistical test help in determining whether or not afforestation produced a change in the organic content of these soils?

11. For a factory to be located on a flood plain, it should gain more from locational advantages than is lost through flood damage. A factory is built on a flood plain on which flood levels reach an average depth of 3 m (with a standard deviation of 1.5 m). Assuming that all perishable goods are located on the second floor at 5.5 m above the flood plain, and given that profits are such that this firm can stand losses in 1 year in 5, how could a decision be made on whether or not investment in this firm is well founded?

12. The height of a river terrace was measured at 25 m intervals in a straight line parallel to the present course of the river. In theory such terraces have a regular, smooth slope, but in practice irregularities are common. What technique could be used to approximate the theoretical surface?

13. The following figures are concerned with the distribution of springs in a region with three different rock types:

Rock type	No. of springs	% of the area
Calcareous marl	18	45
Limestone	26	32
Sandstone	5	23

How can a test be made of the proposition that the frequency of springs is controlled by rock type?

14. Von Thunen's theory concerns the zonation of land uses around a city market. How could the following data be used to test whether or not land uses fall into zones based on distance from the city:

Distance from the city centre (km)	No. of farms according to land use			
	Horticulture	Crops	Dairying	Beef
0–24.9	50	20	20	10
25–89.9	40	110	20	30
90–190	15	10	60	15
> 190	50	10	20	70

Further Reading

THERE is a large number of textbooks on quantitative and statistical techniques written specifically for use by geographers. These should be consulted for further information on the various techniques included in the manual and for an introduction to other and more advanced techniques.

A relatively simple treatment is given in:
1. HAMMOND, P. and McCULLAGH, P. S. (1978) *Quantitative Techniques in Geography: An Introduction.* Oxford University Press, Oxford.

A more advanced text, giving a full discussion of statistical principles and many short exercises which give an indication of the range of possible applications is:
2. SILK, J. (1979) *Statistical Concepts in Geography.* George Allen & Unwin, London.

A thorough treatment of inferential statistics with interesting comments on applications in the geographical literature is given in:
3. NORCLIFFE, G. B. (1977) *Inferential Statistics for Geographers.* Hutchinson, London.

A broader view of quantitative approaches to Geography is taken by:
4. TAYLOR, P. J. (1977) *Quantitative Techniques in Geography: An Introduction to Spatial Analysis.* Houghton Mifflin, Boston.

A formal treatment with many techniques not included in the other texts, but rather difficult reading, is found in:
5. LEWIS, P. (1977) *Maps and Statistics.* Methuen, London.

Clearly presented worked examples are given in:
6. EBDON, D. (1977) *Statistics in Geography: A Practical Approach.* Blackwell, Oxford.

The earliest introductory textbook on statistics in Geography, which has been revised and still has much to recommend it is:
7. GREGORY, S. (1978) *Statistical Methods and the Geographer.* Longman, London.

A readable account, with useful examples, but limited to Human Geography is:
8. SMITH, D. M. (1975) *Patterns in Human Geography.* Penguin Books, Harmondsworth, Middlesex.

References

AIKEN, S. R. (1978) Dengue haemorrhagic fever in south-east Asia. *Transactions of the Institute of British Geographers*, New Series, 3, pp. 476–497.

BERGTHÓRSSON, P. (1969) An estimate of drift ice and temperature in Iceland in 1000 years. *Jökull*, 19, pp. 94–101.

BERRY, B. J. L., SIMMONS, J. W. and TENNANT, R. J. (1963) Urban population densities: structure and change. *Geographical Review*, 53, pp. 389–405.

BLALOCK, H. M. (1960) *Social Statistics*. McGraw-Hill, New York.

BROSSET, C. (1973) Air-borne acid. *Ambio*, 2, pp. 2–9.

CHUNG, R. (1970) Space-time diffusion of the transition model: the twentieth century patterns. In *Population Geography: A Reader* (edited by G. L. Demko, H. M. Rose and G. A. Schnell). McGraw-Hill, New York, pp. 220–239.

CLARK, C. (1951) Urban population densities. *Journal of the Royal Statistical Society*, 114, pp. 490–496.

DAWSON, A. (1979) Raised Shorelines of Jura, Scarba and N.E. Islay. Ph.D. Thesis, University of Edinburgh (Dept. of Geography).

DICKS, T. R. B. (1972) Network analysis and historical geography. *Area*, 4, pp. 4–9. (See also comments and reply in *Area*, 4, pp. 137–141, 279–280.)

DRAKE, M. (1965) The growth of population in Norway 1735–1855. *Scandinavian Economic History Review*, 13, pp. 97–142.

EVANS, I. S. (1977) World-wide variations in the direction and concentration of cirque and glacier aspects. *Geografiska Annaler*, 59(A), pp. 151–175.

FISHER, R. A. and YATES, F. (1974) *Statistical Tables for Biological, Agricultural and Medical Research*. Longman, London.

GONZALÉS, A. (1967) Some effects of population growth on Latin America's economy. *Journal of Inter-American Studies*, 9, pp. 22–42.

GOULD, P. (1970) Is *Statistix inferens* the geographical name for a wild goose? *Economic Geography*, 46 (Supplement), pp. 439–448.

GREGORY, S. (1957) Annual rainfall probability maps of the British Isles. *Quarterly Journal of the Royal Meteorological Society*, 83, pp. 543–549.

GREGORY, S. (1978) *Statistical Methods and the Geographer*. Longman, London.

GRIFFITHS, J. C. (1967) *Scientific Method in Analysis of Sediments*. McGraw-Hill, New York.

HARVEY, D. (1969) *Explanation in Geography*. Arnold, London.

HOLDGATE, M. W. and WHITE, G. F. (1977) *Environmental Issues* (SCOPE Report 10). Wiley, New York.

JACOBS, J. D. and SABO, G. III (1978) Environments and adaptations of the Thule culture on the Davis Strait coast of Baffin Island. *Arctic and Alpine Research*, 10, pp. 595–615.

JONES, P. N. (1978) The distribution and diffusion of the coloured population in England and Wales, 1961–71. *Transactions of the Institute of British Geographers*, New Series, 3, pp. 515–532.

KARLÉN, W. (1973) Holocene glacier and climatic variations, Kebnekaise Mountains, Swedish Lappland. *Geografiska Annaler*, 55(A), pp. 29–63.

LAYTON, R. L. (1978) The operational structure of the hobby farm. *Area*, 10, pp. 242–246.

LEWIS, P. (1977) *Maps and Statistics*. Methuen, London.

LINDLEY, D. V. and MILLER, J. C. P. (1966) *Cambridge Elementary Statistical Tables*. Cambridge University Press, Cambridge.

MATTHEWS, J. A. (1976) 'Little Ice Age' palaeotemperatures from high altitude tree growth in S. Norway. *Nature*, 264, pp. 243–245.

MATTHEWS, J. A. (1977) A lichenometric test of the 1750 end-moraine hypothesis: Storbreen gletschervorfeld, southern Norway. *Norsk Geografisk Tidsskrift*, 31, pp. 129–136.

MATTHEWS, J. A. (1978) Plant colonisation patterns on a gletschervorfeld, southern Norway: a meso-scale geographical approach to vegetation change and phytometric dating. *Boreas*, 7, pp. 155–178.

MATTHEWS, J. A., CORNISH, R. and SHAKESBY, R. A. (1979) 'Saw-tooth' moraines in front of Bödalsbreen, southern Norway. *Journal of Glaciology*, 22, pp. 535–546.

MITCHELL, B. A. (1971) A comparison of chi-square and Kolmogorov–Smirnov tests. *Area*, 3, pp. 237–241.

MORONEY, M. J. (1956) *Facts from Figures*. Penguin Books, Harmondsworth, Middlesex.

NEAVE, H. R. (1978) *Statistical Tables for Mathematicians, Engineers, Economists and the Behavioural and Management Sciences*. George Allen & Unwin, London.

NORCLIFFE, G. B. (1977) *Inferential Statistics for Geographers*. Hutchinson, London.

NRIAGU, J. O. and HARVEY, H. H. (1978) Isotopic variation as an index of sulphur pollution in lakes around Sudbury, Ontario. *Nature*, **273**, pp. 223–224.

Ordnance Survey (1960) *Rhyll, Sheet 95*. Map prepared by the Soil Survey of England and Wales. Ordnance Survey, Chessington, Surrey.

Ordnance Survey (1967) *Rainfall. Annual Average 1916–1950*. Two map sheets prepared by the Ministry of Housing and Local Government. Ordnance Survey, Chessington, Surrey.

Ordnance Survey (1970) *Rhyll, Sheet 95*. Map prepared by the Geological Survey of Great Britain (England and Wales). Ordnance Survey, Chessington, Surrey.

ROBINSON, A. H. and BRYSON, R. A. (1957) A method for describing quantitatively the correspondence of geographical distributions. *Annals of the Association of American Geographers*, **47**, pp. 379–391.

RODDA, J. C., DOWNING, R. A. and LAW, F. M. (1976) *Systematic Hydrology*. Newnes–Butterworth, London.

SCHUMM, S. A. (1956) The evolution of the drainage systems and slopes of bandlands at Perth Amboy, New Jersey. *Bulletin of the Geological Society of America*, **67**, pp. 597–646.

SCOTT, P. (1965) The population structure of Australian cities. *Geographical Journal*, **131**, pp. 463–478.

SHAKESBY, R. A. (1977) The Lennoxtown Essexite Erratics Train. Ph.D. Thesis, University of Edinburgh (Dept. of Geography).

SHAKESBY, R. A. and WARD, R. C. A. (1978) Stone shape analysis as a teaching source: a case study. *Classroom Geographer*, October, pp. 3–10.

SIEGEL, S. (1956) *Nonparametric Statistics for the Behavioral Sciences*. McGraw-Hill, New York.

SILK, J. (1979) *Statistical Concepts in Geography*. George Allen & Unwin, London.

SMITH, D. M. (1977) *Human geography: A Welfare Approach*. Arnold, London.

SPIEGELMAN, M. (1965) Mortality trends for causes of death in countries of low mortality. *Demography*, **2**, pp. 115–125.

TAYLOR, P. J. (1977) *Quantitative Methods in Geography: An Introduction to Spatial Analysis*. Houghton Mifflin, Boston.

UNITED NATIONS (1977) *Statistical Yearbook for 1976*. United Nations, New York.

WALLING, D. E. and WEBB, B. W. (1975) Spatial variation of river water quality: a survey of the River Exe. *Transactions of the Institute of British Geographers*, **65**, pp. 155–171.

WALTER, H. (1973) *Vegetation of the Earth in Relation to Climate and the Eco-physiological conditions*. Springer Verlag, Berlin.

WARDLE, P. (1971) An explanation for Alpine timberlines. *New Zealand Journal of Botany*, **9**, pp. 371–402.

ZAR, J. H. (1974) *Biostatistical Analysis*. Prentice-Hall, Englewood Cliffs, New Jersey.

Answers to Numerical Questions

Exercise 4

Q1. Mean   491.4   350.2   411.7   435.5   469.0   363.4   406.9   285.9   255.1   316.2.
    Median 436.0   234.0   421.5   399.5   438.0   333.0   376.5   205.5   220.5   204.0.

Exercise 5

		June	December
Q1. (b)	Median	53.5 mm	91.6
	Quartile deviation	22.8	28.0
(c)	Mean	63.31 mm	93.02
	Standard deviation ($\hat{\sigma}$)	38.33	41.39

Q3. (a) June 60.5 %; December 44.5 %.

Exercise 6

Q1. (a) 84.13 % (or 0.8413).   (e) 1.28.
    (b) 2.275 %.   (f) 1.645.
    (c) 99.730 %.   (g) 0.270 %.
    (d) 81.855 %.

Q2. (a) 72.91 %.   (f) 0.26 mm
    (b) 14.92 %.   (g) 12.92 %.
    (c) 91.62 %.   (h) 12 or 13.
    (d) 56.06 %.   (i) 8.38 %.
    (e) 0.26 mm.

Q3. (a) 3.14 %.   (c) − 99.06
    (b) 9.34 %.   (d) + 61.49

Exercise 7

Ql. (a)	1956	1966
Belfast	50.0 %	69.2 %
Galway	35.8 %	45.2 %
Dublin	44.4 %	62.9 %
Cork	41.3 %	54.4 %
Edinburgh	69.2 %	82.1 %
Manchester	70.5 %	88.1 %
etc.		

Exercise 8

Q1. (a) 1735–39	29.48	(b) 1735–39	21.92
1736–40	29.44	1736–40	23.14
1737–41	28.74	1737–41	27.20
1738–42	27.86	1738–42	32.74
1739–43	27.88	1739–43	33.86
etc.		etc.	

Exercise 9

Q5. (a) $\bar{x}$ 697.517; $\hat{\sigma}$ 871.369.

(b) $\bar{x}$ 2.55160; $\hat{\sigma}$ 0.50635 (using a log-transformation).

Q6. (a) 80.07 dollars.

(b) Min. in richest $10\% = 1584$ dollars.   (c) 93.19%.

Max. in poorest $10\% = 80$ dollars.   (d) 0.0681.

Q7. (a) $\bar{x} = 2.214\,76$ (164 dollars); $\hat{\sigma} = 0.332\,64$ (using a log-transformation).

(b) $\bar{x} = 2.651\,22$ (448 dollars); $\hat{\sigma} = 0.263\,12$.

(c) $\bar{x} = 2.407\,45$ (256 dollars); $\hat{\sigma} = 0.497\,46$.

(d) $\bar{x} = 3.267\,19$ (1850 dollars); $\hat{\sigma} = 0.211\,92$.

Q8. (a) Min. in richest $10\% = 437$ dollars; max. in poorest $10\% = 62$ dollars.

(b) Min. in richest $10\% = 973$ dollars; max. in poorest $10\% = 206$ dollars.

(c) Min. in richest $10\% = 1107$ dollars; max. in poorest $10\% = 59$ dollars.

(d) Min. in richest $10\% = 3454$ dollars; max. in poorest $10\% = 991$ dollars.

Exercise 11

Q1. (a)

Probability level	Sample size			
	10	25	60	∞
90%	1.833	1.711	1.671	1.645
95%	2.262	2.064	2.000	1.960
99%	3.250	2.797	2.660	2.576

(c) All are approximate percentages:

(i) (i) 90% (i) (ii) 10% (i) (iii) 5% (i) (iv) 45%

(ii) (i) 95% (ii) (ii) 5% (ii) (iii) 2.5% (ii) (iv) 47.5%

(iii) (i) 99% (iii) (ii) 1% (iii) (iii) 0.5% (iii) (iv) 49.5%

(d) (i) 1.703.

(ii) 1.703.

(iii) 1.703.

Q2. (a) 63.31 ± 16.998 mm.

(b) Max. $= 111.375$ mm; min. $= 74.665$ mm.

(d) June 63.31 ± 14.064 mm.

December 93.02 ± 15.187 mm.

(f) Less than 99.9% but greater than 99.0%.

(g) $< 0.5\%$, $> 0.05\%$.

Q3. (a) ±140 ±109 ±73 ±78 ±96 ±79 ±115 ±78 ±50 ±128.

(c) $n = 234$.

Exercise 12

Q1. (a)   5.88   12.80   13.88   17.18   14.49   13.15   14.26   15.14   19.56   17.73.

(b)   1.58   4.13   4.51   5.78   4.12   4.11   3.53   5.14   7.45   5.86.

(c)   0.41   0.77   0.81   1.11   0.58   0.82   0.57   0.99   1.34   0.93.

(d) ±0.87 ±1.57 ±1.65 ±2.28 ±1.17 ±1.69 ±1.16 ±2.03 ±2.73 ±1.87.

Exercise 13

Answers are given for Farndale.

Q1. (b) N&E (includes NW): $\bar{x} = 709.375$; $\hat{\sigma} = 57.6447$.

S&W (includes SE): $\bar{x} = 795.606$; $\hat{\sigma} = 119.8681$.

(c) At 5% significance level, with 47 degrees of freedom, and calculated t statistic of 3.40, the hypothesis of 'no difference' is rejected.

Q2. (a) (i) $t = 13.228$ (with 15 degrees of freedom and a 5% significance level, the hypothesis of 'no difference' is rejected.)

(ii) $t = 5.003$ (32 degrees of freedom).

(iii) $t = 8.371$ (48 degrees of freedom).

(b) $> 99.9\%$ in all cases.

Q3. (a) $\bar{x}_D = 5.8776$; $s_D = 6.6964$.

(b) $t = 6.0810$ (48 degrees of freedom). The difference is significant at the 5% significance level.

(c) $< 5\%$.

Q4. Also significant at the 5% level (use the 10% column of Table C).

Exercise 14

Q3. (a) $\chi^2 = 76.45$ (degrees of freedom $= 3$). The difference is significant at the 5% level (and all tabulated significance levels).

(b) $\chi^2 = 35.89$ (degrees of freedom $= 1$). The difference is significant at the 5% level (and all tabulated significance levels).

(c) $\chi^2 = 7.44$ (degrees of freedom $= 2$). The difference is significant at the 5% level (but not at the 1% level).

Q4. (a) After combining N and E aspects to make expected frequencies sufficiently large, $\chi^2 = 1.523$ (degrees of freedom $= 2$ for a 3×2 contingency table). Cannot reject the hypothesis of 'no difference' at the 5% level (or 1% level).

(b) $\chi^2 = 2.05$ (with 1 degree of freedom for a 2×2 contingency table).

(c) $\chi^2 = 3.949$ (with 2 degrees of freedom for a 3×2 contingency table).

Q5. (a) Test not valid because two cells have expected frequencies below 5.

(b) Test not valid because two cells have expected frequencies below 5.

(c) $\chi^2 = 7.943$ (degrees of freedom $= 2$ for a 3×2 contingency table). The hypothesis of 'no difference' cannot be rejected at the 5% level (only at the 10% level).

Exercise 15

Q1. (a) After combining categories to form two columns, $\chi^2 = 18.70$ (with 2 degrees of freedom for a 2×3 contingency table). The hypothesis of 'no difference' is rejected at the 5% level (and at all the tabulated significance levels).

(b) Woodland versus farmland.
After combining categories to form two columns, $\chi^2 = 0.0174$ (with one degree of freedom for a 2×2 contingency table), the hypothesis of 'no difference' cannot be rejected at the 5% level (or at the 10% level).

(b) Farmland versus moorland.
After combining categories to form two columns, $\chi^2 = 16.70$ (with 1 degree of freedom for a 2×2 contingency table), the hypothesis of 'no difference' is rejected at the 5% level (and at all the tabulated significance levels).

(b) Woodland versus moorland.
$\chi^2 = 21.65$ (with 2 degrees of freedom for a 2×3 contingency table). The hypothesis of 'no difference' is rejected at all tabulated significance levels.

Exercise 16

Q1. (a) $D = 0.31$ (tabulated $D = 0.1822$ at the 5% level of significance). The hypothesis of 'no difference' is rejected at the 5% significance level (and at all tabulated significance levels).

(b) $D = 0.41$. The hypothesis of 'no difference' is rejected at all the tabulated significance levels.

(c) $D = 0.28$. The hypothesis of 'no difference' is rejected at all the tabulated significance levels.

Exercise 17

Q1. $H = 68.2$. Using Table D, with 3 degrees of freedom, the hypothesis of 'no difference' is rejected at the 5% level (and at all the tabulated significance levels).

Exercise 18

Q1. (a) Teeth 4.06; notches 6.73.   (c) Teeth 10.72; notches 19.41.
(b) Teeth 38.00; notches 41.05.   (d) Teeth 19.22; notches 22.55.

Q3. (a) $U = 84$ (a significant difference at the 5% and 1% levels).
(b) $U = 73.5$ (not a significant difference at the 5% level, only at the 10% significance level).
(c) $U = 63.5$ (not a significant difference at the 5% level, or at the 10% level).
(d) $U = 86$ (a significant difference at the 5% and 1% levels).

Q5. (a) $T = 3$ (a significant difference at the 5% and 2% levels).
(b) $T = 14.5$ (not a significant difference at the 5% or 10% levels).

Exercise 19

Q1. (a) $\chi_r^2 = 16.9$. Using Table D with 2 degrees of freedom, the difference is significant at the 5% level (and at all tabulated significance levels).

Q3. (b) $\chi_r^2 = 98.9$. Using Table D with 8 degrees of freedom, the difference is significant at the 5% level (and at all tabulated significance levels).

Exercise 20

Q3. (a) $r = -0.73$ ($t = 7.477$ with 49 degrees of freedom). The relationship is significant at the 5% level.

(b) $r = +0.21$ ($t = 1.4568$ with 46 degrees of freedom). The relationship is not significant at the 5% level (only at the 20% significance level).

(c) $r = +0.55$ ($t = 4.4665$ with 46 degrees of freedom). The relationship is significant at the 5% level.

Exercise 21

Q2. (i) $r_s = +0.754$ (significant at the 5% level and at all tabulated significance levels).

(ii) $r_s = -0.708$ (significant at the 5% level and also at the 1% level).

(iii) $r_s = +0.505$ (not significant at the 5% level, only at the 10% level with $n = 12$).

(iv) $r_s = +0.305$ (not significant at the 5% level, or at the 10% level).

Exercise 22

Q1. (c) $r = -0.5945$.

(d) $\hat{y} = 3.5458 - 0.2547\,x$.

Q2. (a) $t = 4.9596$ (significant at all tabulated significance levels).

(b) $r^2 = 0.3534$ (or 35.34%).

Q3. (a) 2.02°C.

(b) 3.55°C (a coefficient).

(c) -1.53°C (b coefficient).

Appendix: Statistical Tables

TABLE A. *Tables of the z statistic (the normal distribution function)*

z	p	z	p	z	p	z	p	z	p	z	p
0.00	0.5000	0.50	0.6915	1.00	0.8413	1.50	0.9332	2.00	0.97725	2.50	0.99379
0.01	0.5040	0.51	0.6950	1.01	0.8438	1.51	0.9345	2.01	0.97778	2.51	0.99396
0.02	0.5080	0.52	0.6985	1.02	0.8461	1.52	0.9357	2.02	0.97831	2.52	0.99413
0.03	0.5120	0.53	0.7019	1.03	0.8485	1.53	0.9370	2.03	0.97882	2.53	0.99430
0.04	0.5160	0.54	0.7054	1.04	0.8508	1.54	0.9382	2.04	0.97932	2.54	0.99446
0.05	0.5199	0.55	0.7088	1.05	0.8531	1.55	0.9394	2.05	0.97982	2.55	0.99461
0.06	0.5239	0.56	0.7123	1.06	0.8554	1.56	0.9406	2.06	0.98030	2.56	0.99477
0.07	0.5279	0.57	0.7157	1.07	0.8577	1.57	0.9418	2.07	0.98077	2.57	0.99492
0.08	0.5319	0.58	0.7190	1.08	0.8599	1.58	0.9429	2.08	0.98124	2.58	0.99506
0.09	0.5359	0.59	0.7224	1.09	0.8621	1.59	0.9441	2.09	0.98169	2.59	0.99520
0.10	0.5398	0.60	0.7257	1.10	0.8643	1.60	0.9452	2.10	0.98214	2.60	0.99534
0.11	0.5438	0.61	0.7291	1.11	0.8665	1.61	0.9463	2.11	0.98257	2.61	0.99547
0.12	0.5478	0.62	0.7324	1.12	0.8686	1.62	0.9474	2.12	0.98300	2.62	0.99560
0.13	0.5517	0.63	0.7357	1.13	0.8708	1.63	0.9484	2.13	0.98341	2.63	0.99573
0.14	0.5557	0.64	0.7389	1.14	0.8729	1.64	0.9495	2.14	0.98382	2.64	0.99585
0.15	0.5596	0.65	0.7422	1.15	0.8749	1.65	0.9505	2.15	0.98422	2.65	0.99598
0.16	0.5636	0.66	0.7454	1.16	0.8770	1.66	0.9515	2.16	0.98461	2.66	0.99609
0.17	0.5675	0.67	0.7486	1.17	0.8790	1.67	0.9525	2.17	0.98500	2.67	0.99621
0.18	0.5714	0.68	0.7517	1.18	0.8810	1.68	0.9535	2.18	0.98537	2.68	0.99632
0.19	0.5753	0.69	0.7549	1.19	0.8830	1.69	0.9545	2.19	0.98574	2.69	0.99643
0.20	0.5793	0.70	0.7580	1.20	0.8849	1.70	0.9554	2.20	0.98610	2.70	0.99653
0.21	0.5832	0.71	0.7611	1.21	0.8869	1.71	0.9564	2.21	0.98645	2.71	0.99664
0.22	0.5871	0.72	0.7642	1.22	0.8888	1.72	0.9573	2.22	0.98679	2.72	0.99674
0.23	0.5910	0.73	0.7673	1.23	0.8907	1.73	0.9582	2.23	0.98713	2.73	0.99683
0.24	0.5948	0.74	0.7704	1.24	0.8925	1.74	0.9591	2.24	0.98745	2.74	0.99693
0.25	0.5987	0.75	0.7734	1.25	0.8944	1.75	0.9599	2.25	0.98778	2.75	0.99702
0.26	0.6026	0.76	0.7764	1.26	0.8962	1.76	0.9608	2.26	0.98809	2.76	0.99711
0.27	0.6064	0.77	0.7794	1.27	0.8980	1.77	0.9616	2.27	0.98840	2.77	0.99720
0.28	0.6103	0.78	0.7823	1.28	0.8997	1.78	0.9625	2.28	0.98870	2.78	0.99728
0.29	0.6141	0.79	0.7852	1.29	0.9015	1.79	0.9633	2.29	0.98899	2.79	0.99736
0.30	0.6179	0.80	0.7881	1.30	0.9032	1.80	0.9641	2.30	0.98928	2.80	0.99744
0.31	0.6217	0.81	0.7910	1.31	0.9049	1.81	0.9649	2.31	0.98956	2.81	0.99752
0.32	0.6255	0.82	0.7939	1.32	0.9066	1.82	0.9656	2.32	0.98983	2.82	0.99760
0.33	0.6293	0.83	0.7967	1.33	0.9082	1.83	0.9664	2.33	0.99010	2.83	0.99767
0.34	0.6331	0.84	0.7995	1.34	0.9099	1.84	0.9671	2.34	0.99036	2.84	0.99774
0.35	0.6368	0.85	0.8023	1.35	0.9115	1.85	0.9678	2.35	0.99061	2.85	0.99781
0.36	0.6406	0.86	0.8051	1.36	0.9131	1.86	0.9686	2.36	0.99086		
0.37	0.6443	0.87	0.8078	1.37	0.9147	1.87	0.9693	2.37	0.99111	2.90	0.99813
0.38	0.6480	0.88	0.8106	1.38	0.9162	1.88	0.9699	2.38	0.99134		
0.39	0.6517	0.89	0.8133	1.39	0.9177	1.89	0.9706	2.39	0.99158	2.95	0.99841
0.40	0.6554	0.90	0.8159	1.40	0.9192	1.90	0.9713	2.40	0.99180	3.00	0.99865
0.41	0.6591	0.91	0.8186	1.41	0.9207	1.91	0.9719	2.41	0.99202	3.10	0.99903
0.42	0.6628	0.92	0.8212	1.42	0.9222	1.92	0.9726	2.42	0.99224	3.20	0.99931
0.43	0.6664	0.93	0.8238	1.43	0.9236	1.93	0.9732	2.43	0.99245	3.30	0.99952
0.44	0.6700	0.94	0.8264	1.44	0.9251	1.94	0.9738	2.44	0.99266	3.40	0.99966
0.45	0.6736	0.95	0.8289	1.45	0.9265	1.95	0.9744	2.45	0.99286	3.50	0.99977
0.46	0.6772	0.96	0.8315	1.46	0.9279	1.96	0.9750	2.46	0.99305	3.60	0.99984
0.47	0.6808	0.97	0.8340	1.47	0.9292	1.97	0.9756	2.47	0.99324	3.70	0.99989
0.48	0.6844	0.98	0.8365	1.48	0.9306	1.98	0.9761	2.48	0.99343	3.80	0.99993
0.49	0.6879	0.99	0.8389	1.49	0.9319	1.99	0.9767	2.49	0.99361	3.90	0.99995
0.50	0.6915	1.00	0.8413	1.50	0.9332	2.00	0.9772	2.50	0.99379	4.00	0.99997

(From Lindley and Miller, 1966.)

TABLE B. *Tables of random sampling numbers*

57780	97609	52482	12783	88768	12323	64967	22970	11204	37576
68327	00067	17487	49149	25894	23639	86557	04139	10756	76285
55888	82253	67464	91628	88764	43598	45481	00331	15900	97699
84910	44827	31173	44247	56573	91759	79931	26644	27048	53704
35654	53638	00563	57230	07395	10813	99194	81592	96834	21374
46381	60071	20835	43110	31842	02855	73446	24456	24268	85291
11212	06034	77313	66896	47902	63483	09924	83635	30013	61791
49703	07226	73337	49223	73312	09534	64005	79267	76590	26066
05482	30340	24606	99042	16536	14267	84084	16198	94852	44305
92947	65090	47455	90675	89921	13036	92867	04786	76776	18675
51806	61445	32437	01129	03644	70024	07629	55805	85616	59569
16383	30577	91319	67998	72423	81307	75192	80443	09651	30068
30893	85406	42369	71836	74479	68273	78133	34506	68711	58725
59790	11682	63156	10443	99033	76460	36814	36917	37232	66218
06271	74980	46094	21881	43525	16516	26393	89082	24343	57546
93325	61834	40763	81178	17507	90432	50973	35591	36930	03184
46690	08927	32962	24882	83156	58597	88267	32479	80440	41668
82041	88942	57572	34539	43812	58483	43779	42718	46798	49079
14306	04003	91186	70093	62700	99408	72236	52722	37531	24590
63471	77583	80056	59027	37031	05819	90836	19530	07138	36431
68467	17634	84211	31776	92996	75644	82043	84157	10877	12536
94308	57895	08121	07088	65080	51928	74237	00449	86625	06626
52218	32502	82195	43867	79935	34620	37386	00243	46353	44499
46586	08309	52702	85464	06670	18796	74713	81632	34056	56461
07869	80471	69139	82408	33989	44250	79597	15182	14956	70423
46719	60281	88638	26909	32415	31864	53708	60219	44482	40004
74687	71227	59716	80619	56816	73807	94150	21991	22901	74351
42731	50249	11685	54034	12710	35159	00214	19440	61539	25717
71740	29429	86822	01187	96497	25823	18415	06087	05886	11205
96746	05938	11828	47727	02522	33147	92846	15010	96725	67903
27564	81744	51909	36192	45263	33212	71808	24753	72644	74441
21895	29683	26533	14740	94286	90342	24674	52762	22051	31743
01492	40778	05988	65760	13468	31132	37106	02723	40202	15824
55846	19271	22846	80425	00235	34292	72181	24910	25245	81239
14615	75196	40313	50783	66585	39010	76796	31385	26785	66830
77848	15755	91938	81915	65312	86956	26195	61525	97406	67988
87167	03106	52876	31670	23850	13257	77510	42393	53782	32412
73018	56511	89388	73133	12074	62538	57215	23476	92150	14737
29247	67792	10593	22772	03407	24319	19525	24672	21182	10765
17412	09161	34905	44524	20124	85151	25952	81930	43536	39705
68805	19830	87973	99691	25096	41497	57562	35553	77057	06161
40551	36740	61851	76158	35441	66188	87728	66375	98049	84604
90379	06314	21897	42800	63963	44258	14381	90884	66620	14538
09466	65311	95514	51559	29960	07521	42180	86677	94240	59783
15821	25078	19388	93798	50820	88254	20504	74158	35756	42100
10328	60890	05204	30069	79630	31572	63273	13703	52954	72793
49727	08160	81650	71690	56327	06729	22495	49756	43333	34533
71118	41798	34541	76432	40522	51521	74382	06305	11956	30611
53253	23100	03743	48999	37736	92186	19108	69017	21661	17175
12206	24205	32372	46438	67981	53226	24943	68659	91924	69555

(From Neave, 1978.)

TABLE C. *Tables of the Student t statistic*

Degrees of freedom	Significance level					
	20%	10%	5%	2%	1%	0.1%
1	3.078	6.314	12.706	31.821	63.657	636.619
2	1.886	2.920	4.303	6.965	9.925	31.598
3	1.638	2.353	3.182	4.541	5.841	12.941
4	1.533	2.132	2.776	3.747	4.604	8.610
5	1.476	2.015	2.571	3.365	4.032	6.869
6	1.440	1.943	2.447	3.143	3.707	5.959
7	1.415	1.895	2.365	2.998	3.499	5.408
8	1.397	1.860	2.306	2.896	3.355	5.041
9	1.383	1.833	2.262	2.821	3.250	4.781
10	1.372	1.812	2.228	2.764	3.169	4.587
11	1.363	1.796	2.201	2.718	3.106	4.437
12	1.356	1.782	2.179	2.681	3.055	4.318
13	1.350	1.771	2.160	2.650	3.012	4.221
14	1.345	1.761	2.145	2.624	2.977	4.140
15	1.341	1.753	2.131	2.602	2.947	4.073
16	1.337	1.746	2.120	2.583	2.921	4.015
17	1.333	1.740	2.110	2.567	2.898	3.965
18	1.330	1.734	2.101	2.552	2.878	3.922
19	1.328	1.729	2.093	2.539	2.861	3.883
20	1.325	1.725	2.086	2.528	2.845	3.850
21	1.323	1.721	2.080	2.518	2.831	3.819
22	1.321	1.717	2.074	2.508	2.819	3.792
23	1.319	1.714	2.069	2.500	2.807	3.768
24	1.318	1.711	2.064	2.492	2.797	3.745
25	1.316	1.708	2.060	2.485	2.787	3.725
26	1.315	1.706	2.056	2.479	2.779	3.707
27	1.314	1.703	2.052	2.473	2.771	3.690
28	1.313	1.701	2.048	2.467	2.763	3.674
29	1.311	1.699	2.045	2.462	2.756	3.659
30	1.310	1.697	2.042	2.457	2.750	3.646
40	1.303	1.684	2.021	2.423	2.704	3.551
50	1.299	1.676	2.009	2.403	2.678	3.496
60	1.296	1.671	2.000	2.390	2.660	3.460
70	1.294	1.667	1.994	2.381	2.648	3.435
80	1.292	1.664	1.990	2.374	2.639	3.416
90	1.291	1.662	1.987	2.368	2.632	3.402
100	1.290	1.660	1.984	2.364	2.626	3.390
120	1.289	1.658	1.980	2.358	2.617	3.373
150	1.287	1.655	1.976	2.351	2.609	3.357
∞	1.282	1.645	1.960	2.326	2.576	3.291

(From Fisher and Yates, 1974)

TABLE D. *Tables of the χ² statistic*

Degrees of Freedom	Significance level									
	99.5%	99%	97.5%	95%	10%	5%	2.5%	1%	0.5%	0.1%
1	0.0000393	0.000157	0.000982	0.00393	2.71	3.84	5.02	6.63	7.88	10.83
2	0.0100	0.0201	0.0506	0.103	4.61	5.99	7.38	9.21	10.60	13.81
3	0.0717	0.115	0.216	0.352	6.25	7.81	9.35	11.35	12.84	16.27
4	0.207	0.297	0.484	0.711	7.78	9.49	11.14	13.28	14.86	18.47
5	0.412	0.554	0.831	1.15	9.24	11.07	12.83	15.09	16.75	20.52
6	0.676	0.872	1.24	1.64	10.64	12.59	14.45	16.81	18.55	22.46
7	0.989	1.24	1.69	2.17	12.02	14.07	16.01	18.48	20.28	24.32
8	1.34	1.65	2.18	2.73	13.36	15.51	17.53	20.09	21.95	26.12
9	1.73	2.09	2.70	3.33	14.68	16.92	19.02	21.67	23.59	27.88
10	2.16	2.56	3.25	3.94	15.99	18.31	20.48	23.21	25.19	29.59
11	2.60	3.05	3.82	4.57	17.28	19.68	21.92	24.73	26.76	31.26
12	3.07	3.57	4.40	5.23	18.55	21.03	23.34	26.22	28.30	32.91
13	3.57	4.11	5.01	5.89	19.81	22.36	24.74	27.69	29.82	34.53
14	4.07	4.66	5.63	6.57	21.06	23.68	26.12	29.14	31.32	36.12
15	4.60	5.23	6.26	7.26	22.31	25.00	27.49	30.58	32.80	37.70
16	5.14	5.81	6.91	7.96	23.54	26.30	28.85	32.00	34.27	39.25
17	5.70	6.41	7.56	8.67	24.77	27.59	30.19	33.41	35.72	40.79
18	6.26	7.01	8.23	9.39	25.99	28.87	31.53	34.81	37.16	42.31
19	6.84	7.63	8.91	10.12	27.20	30.14	32.85	36.19	38.58	43.82
20	7.43	8.26	9.59	10.85	28.41	31.41	34.17	37.57	40.00	45.31
21	8.03	8.90	10.28	11.59	29.62	32.67	35.48	38.93	41.40	46.80
22	8.64	9.54	10.98	12.34	30.81	33.92	36.78	40.29	42.80	48.27
23	9.26	10.20	11.69	13.09	32.01	35.17	38.08	41.64	44.18	49.73
24	9.89	10.86	12.40	13.85	33.20	36.42	39.36	42.98	45.56	51.18
25	10.52	11.52	13.12	14.61	34.38	37.65	40.65	44.31	46.93	52.62
26	11.16	12.20	13.84	15.38	35.56	38.89	41.92	45.64	48.29	54.05
27	11.81	12.88	14.57	16.15	36.74	40.11	43.19	46.96	49.64	55.48
28	12.46	13.56	15.31	16.93	37.92	41.34	44.46	48.28	50.99	56.89
29	13.12	14.26	16.05	17.71	39.09	42.56	45.72	49.59	52.34	58.30
30	13.79	14.95	16.79	18.49	40.26	43.77	46.98	50.89	53.67	59.70
40	20.71	22.16	24.43	26.51	51.81	55.76	59.34	63.69	66.77	73.40
50	27.99	29.71	32.36	34.76	63.17	67.50	71.42	76.15	79.49	86.66
60	35.53	37.48	40.48	43.19	74.40	79.08	83.30	88.38	91.95	99.61
70	43.28	45.44	48.76	51.74	85.53	90.53	95.02	100.4	104.2	112.3
80	51.17	53.54	57.15	60.39	96.58	101.9	106.6	112.3	116.3	124.8
90	59.20	61.75	65.65	69.13	107.6	113.1	118.1	124.1	128.3	137.2
100	67.33	70.06	74.22	77.93	118.5	124.3	129.6	135.8	140.2	149.4
120	83.85	86.92	91.57	95.70	140.2	146.6	152.2	159.0	163.6	173.6
150	109.1	112.7	118.0	122.7	172.6	179.6	185.8	193.2	198.4	209.3
200	152.2	156.4	162.7	168.3	226.0	234.0	241.1	249.4	255.3	267.5

(From Lindley and Miller, 1966, with additions from Neave, 1978.)

TABLE E. *Tables of the Kolmogorov–Smirnov D statistic*

Sample size (n)	Significance level 5%	Significance level 1%
3	—	—
4	1.0000	—
5	1.0000	1.0000
6	0.8333	1.0000
7	0.8571	0.8571
8	0.7500	0.8750
9	0.6667	0.7778
10	0.7000	0.8000
11	0.6364	0.7273
12	0.5833	0.6667
13	0.5385	0.6923
14	0.5714	0.6429
15	0.5333	0.6000
16	0.5000	0.6250
17	0.4706	0.5882
18	0.5000	0.5556
19	0.4737	0.5263
20	0.4500	0.5500
21	0.4286	0.5238
22	0.4091	0.5000
23	0.4348	0.4783
24	0.4167	0.5000
25	0.4000	0.4800
26	0.3846	0.4615
27	0.3704	0.4444
28	0.3929	0.4643
29	0.3793	0.4483
30	0.3667	0.4333
35	0.3429	—
40	0.3250	—

	5%	1%
> 40	$1.36\sqrt{\dfrac{n_1 + n_2}{n_1 n_2}}$	$1.63\sqrt{\dfrac{n_1 + n_2}{n_1 n_2}}$

	10%	0.1%
> 40	$1.22\sqrt{\dfrac{n_1 + n_2}{n_1 n_2}}$	$1.95\sqrt{\dfrac{n_1 + n_2}{n_1 n_2}}$

(From Taylor, 1977.)

TABLE F. *Tables of the Mann-Whitney U statistic*

Sample sizes		Significance level				
n_1	n_2	10%	5%	2%	1%	0.1%
1	1–15	–	–	–	–	–
	16	–	–	–	–	–
	17	–	–	–	–	–
	18	–	–	–	–	–
	19	19	–	–	–	–
	20	20	–	–	–	–
	30	30	–	–	–	–
2	2	–	–	–	–	–
	3	–	–	–	–	–
	4	–	–	–	–	–
	5	10	–	–	–	–
	6	12	–	–	–	–
	7	14	–	–	–	–
	8	15	16	–	–	–
	9	17	18	–	–	–
	10	19	20	–	–	–
	11	21	22	–	–	–
	12	22	23	–	–	–
	13	24	25	26	–	–
	14	25	27	28	–	–
	15	27	29	30	–	–
	16	29	31	32	–	–
	17	31	32	34	–	–
	18	32	34	36	–	–
	19	34	36	37	38	–
	20	36	38	39	40	–
	30	53	55	58	59	–
3	3	9	–	–	–	–
	4	12	–	–	–	–
	5	14	15	–	–	–
	6	16	17	–	–	–
	7	19	20	21	–	–
	8	21	22	24	–	–
	9	23	25	26	27	–
	10	26	27	29	30	–
	11	28	30	32	33	–
	12	31	32	34	35	–
	13	33	35	37	38	–
	14	35	37	40	41	–
	15	38	40	42	43	–
	16	40	42	45	46	–
	17	42	45	47	49	–
	18	45	47	50	52	–
	19	47	50	53	54	–
	20	49	52	55	57	–
	30	73	77	81	84	89

TABLE F. (*cont.*)

Sample sizes		Significance level				
n_1	n_2	10%	5%	2%	1%	0.1%
4	4	15	16	–	–	–
	5	18	19	20	–	–
	6	21	22	23	24	–
	7	24	25	27	28	–
	8	27	28	30	31	–
	9	30	32	33	35	–
	10	33	35	37	38	–
	11	36	38	40	42	–
	12	39	41	43	45	–
	13	42	44	47	49	52
	14	45	47	50	52	56
	15	48	50	53	55	60
	16	50	53	57	59	63
	17	53	57	60	62	67
	18	56	60	63	66	71
	19	59	63	67	69	74
	20	62	66	70	72	78
	30	92	97	103	107	115
5	5	21	23	24	25	–
	6	25	27	28	29	–
	7	29	30	32	34	–
	8	32	34	36	38	–
	9	36	38	40	42	45
	10	39	42	44	46	50
	11	43	46	48	50	54
	12	47	49	52	54	59
	13	50	53	56	58	63
	14	54	57	60	63	68
	15	57	61	64	67	72
	16	61	65	68	71	77
	17	65	68	72	75	81
	18	68	72	76	79	86
	19	72	76	80	83	90
	20	75	80	84	87	95
	30	111	117	124	128	139
6	6	29	31	33	34	–
	7	34	36	38	39	–
	8	38	40	42	44	48
	9	42	44	47	49	53
	10	46	.49	52	54	58
	11	50	53	57	59	64
	12	55	58	61	63	69
	13	59	62	66	68	74
	14	63	67	71	73	79
	15	67	71	75	78	85
	16	71	75	80	83	90
	17	76	80	84	87	95
	18	80	84	89	92	100
	19	84	89	94	97	106
	20	88	93	98	102	111
	30	130	137	145	150	163

TABLE F. (*cont.*)

Sample sizes		Significance level				
n_1	n_2	10%	5%	2%	1%	0.1%
7	7	38	41	43	45	49
	8	43	46	49	50	55
	9	48	51	54	56	61
	10	53	56	59	61	67
	11	58	61	65	67	73
	12	63	66	70	72	79
	13	67	71	75	78	85
	14	72	76	81	83	91
	15	77	81	86	89	97
	16	82	86	91	94	103
	17	86	91	96	100	109
	18	91	96	102	105	115
	19	96	101	107	111	120
	20	101	106	112	116	126
	30	149	156	165	170	185
8	8	49	51	55	57	62
	9	54	57	61	63	68
	10	60	63	67	69	75
	11	65	69	73	75	82
	12	70	74	79	81	89
	13	76	80	84	87	95
	14	81	86	90	94	102
	15	87	91	96	100	109
	16	92	97	102	106	115
	17	97	102	108	112	122
	18	103	108	114	118	129
	19	108	114	120	124	135
	20	113	119	126	130	142
	30	167	175	185	191	208
9	9	60	64	67	70	76
	10	66	70	74	77	83
	11	72	76	81	83	91
	12	78	82	87	90	98
	13	84	89	94	97	106
	14	90	95	100	104	113
	15	96	101	107	111	120
	16	102	107	113	117	128
	17	108	114	120	124	135
	18	114	120	126	131	142
	19	120	126	133	138	150
	20	126	132	140	144	157
	30	185	194	205	212	230
10	10	73	77	81	84	92
	11	79	84	88	92	100
	12	86	91	96	99	108
	13	93	97	103	106	116
	14	99	104	110	114	124
	15	106	111	117	121	132
	16	112	118	124	129	140
	17	119	125	132	136	148
	18	125	132	139	143	156
	19	132	138	146	151	164
	20	138	145	153	158	172
	30	204	213	224	232	252

TABLE F. (*cont.*)

Sample sizes		Significance level				
n_1	n_2	10 %	5 %	2 %	1 %	0.1 %
11	11	87	91	96	100	109
	12	94	99	104	108	117
	13	101	106	112	116	126
	14	108	114	120	124	135
	15	115	121	128	132	144
	16	122	129	135	140	152
	17	130	136	143	148	161
	18	137	143	151	156	170
	19	144	151	159	164	178
	20	151	158	167	172	187
	30	222	232	244	252	273
12	12	102	107	113	117	127
	13	109	115	121	125	136
	14	117	123	130	134	146
	15	125	131	138	143	155
	16	132	139	146	151	165
	17	140	147	155	160	174
	18	148	155	163	169	183
	19	156	163	172	177	193
	20	163	171	180	186	202
	30	240	251	264	272	295
13	13	118	124	130	135	146
	14	126	132	139	144	157
	15	134	141	148	153	167
	16	143	149	157	163	177
	17	151	158	166	172	187
	18	159	167	175	181	197
	19	167	175	184	190	207
	20	176	184	193	200	217
	30	258	270	283	292	316
14	14	135	141	149	154	167
	15	144	151	159	164	178
	16	153	160	168	174	189
	17	161	169	178	184	199
	18	170	178	187	194	210
	19	179	188	197	203	221
	20	188	197	207	213	231
	30	276	289	302	312	337
15	15	153	161	169	174	189
	16	163	170	179	185	201
	17	172	180	189	195	212
	18	182	190	200	206	224
	19	191	200	210	216	235
	20	200	210	220	227	246
	30	294	307	322	331	358
16	16	173	181	190	196	213
	17	183	191	201	207	225
	18	193	202	212	218	237
	19	203	212	222	230	249
	20	213	222	233	241	261
	30	312	326	341	351	379

TABLE F. (*cont.*)

Sample sizes		Significance level				
n_1	n_2	10%	5%	2%	1%	0.1%
17	17	193	202	212	219	238
	18	204	213	224	231	250
	19	214	224	235	242	263
	20	225	235	247	254	275
	30	330	344	360	371	400
18	18	215	225	236	243	263
	19	226	236	248	255	277
	20	237	248	260	268	287
	30	348	363	379	390	421
19	19	238	248	260	268	291
	20	250	261	273	281	304
	30	366	381	398	410	442
20	20	262	273	286	295	319
	30	384	400	418	430	463

(From Zar, 1974.)

TABLE G. *Tables of the Kruskal–Wallis H statistic*

Sample sizes			Significance level		
n_1	n_2	n_3	10%	5%	1%
3	2	1	4.286	—	—
3	2	2	4.500	4.714	—
3	3	1	4.571	5.143	—
3	3	2	4.556	5.361	—
3	3	3	4.622	5.600	7.200
4	2	1	4.500	—	—
4	2	2	4.458	5.333	—
4	3	1	4.056	5.208	—
4	3	2	4.511	5.444	6.444
4	3	3	4.709	5.727	6.746
4	4	1	4.167	4.967	6.667
4	4	2	4.554	5.455	7.036
4	4	3	4.546	5.598	7.144
4	4	4	4.654	5.692	7.654
5	2	1	4.200	5.000	—
5	2	2	4.373	5.160	6.533
5	3	1	4.018	4.960	—
5	3	2	4.651	5.251	6.909
5	3	3	4.533	5.648	7.079
5	4	1	3.987	4.986	6.954
5	4	2	4.541	5.273	7.204
5	4	3	4.549	5.656	7.445
5	4	4	4.619	5.657	7.760
5	5	1	4.109	5.127	7.309
5	5	2	4.623	5.338	7.338
5	5	3	5.545	5.705	7.578
5	5	4	4.523	5.666	7.823
5	5	5	4.560	5.780	8.000

(From Zar, 1974.)

TABLE H. *Tables of the Wilcoxon T statistic*

Sample size (n)	Significance level				
	10%	5%	2%	1%	0.1%
6	2	–	–	–	–
7	3	2	–	–	–
8	5	3	1	–	–
9	8	5	3	1	–
10	10	8	5	3	–
11	13	10	7	5	–
12	17	13	9	7	1
13	21	17	12	9	2
14	25	21	15	12	4
15	30	25	19	15	6
16	35	29	23	19	8
17	41	34	27	23	11
18	47	40	32	27	14
19	53	46	37	32	18
20	60	52	43	37	21
21	67	58	49	42	25
22	75	65	55	48	30
23	83	73	62	54	35
24	91	81	69	61	40
25	100	89	76	68	45
26	110	98	84	75	51
27	119	107	92	83	57
28	130	116	101	91	64
29	140	126	110	100	71
30	151	137	120	109	78
31	163	147	130	118	86
32	175	159	140	128	94
33	187	170	151	138	102
34	200	182	162	148	111
35	213	195	173	159	120
36	227	208	185	171	130
37	241	221	198	182	140
38	256	235	211	194	150
39	271	249	224	207	161
40	286	264	238	220	172
50	466	434	397	373	304
60	690	648	600	567	476
70	960	907	846	805	689
80	1276	1211	1136	1086	943
90	1638	1560	1471	1410	1240
100	2045	1955	1850	1779	1578

(From Zar, 1974)

TABLE I. *Tables of the Friedman χ_r^2 statistic*

Number of samples (k)	Sample size (n)	Significance level				
		10%	5%	2%	1%	0.1%
3	3	6.000	6.000	–	–	–
3	4	6.000	6.500	8.000	8.000	–
3	5	5.200	6.400	6.400	8.400	10.000
3	6	5.333	7.000	8.333	9.000	12.000
3	7	5.429	7.143	8.000	8.857	12.286
3	8	5.250	6.250	7.750	9.000	12.250
3	9	5.556	6.222	8.000	8.667	12.667
4	2	6.000	6.000	–	–	–
4	3	6.600	7.400	8.200	9.000	–
4	4	6.300	7.800	8.400	9.600	11.100

(From Zar, 1974.)

TABLE J. *Tables of the Spearman rank correlation coefficient*

Sample size (n)	Significance level				
	10%	5%	2%	1%	0.1%
4	1.000	–	–	–	–
5	0.900	1.000	1.000	–	–
6	0.829	0.886	0.943	1.000	–
7	0.714	0.786	0.893	0.929	1.000
8	0.643	0.738	0.833	0.881	0.976
9	0.600	0.700	0.783	0.833	0.933
10	0.564	0.648	0.745	0.794	0.903
11	0.536	0.618	0.709	0.755	0.873
12	0.503	0.587	0.678	0.727	0.846
13	0.484	0.560	0.648	0.703	0.824
14	0.464	0.538	0.626	0.679	0.802
15	0.446	0.521	0.604	0.654	0.779
16	0.429	0.503	0.582	0.635	0.762
17	0.414	0.485	0.566	0.615	0.748
18	0.401	0.472	0.550	0.600	0.728
19	0.391	0.460	0.535	0.584	0.712
20	0.380	0.447	0.520	0.570	0.696
21	0.370	0.435	0.508	0.556	0.681
22	0.361	0.425	0.496	0.544	0.667
23	0.353	0.415	0.486	0.532	0.654
24	0.344	0.406	0.476	0.521	0.642
25	0.337	0.398	0.466	0.511	0.630
26	0.331	0.390	0.457	0.501	0.619
27	0.324	0.382	0.448	0.491	0.608
28	0.317	0.375	0.440	0.483	0.598
29	0.312	0.368	0.433	0.475	0.589
30	0.306	0.362	0.425	0.467	0.580
31	0.301	0.356	0.418	0.459	0.571
32	0.296	0.350	0.412	0.452	0.563
33	0.291	0.345	0.405	0.446	0.554
34	0.287	0.340	0.399	0.439	0.547
35	0.283	0.335	0.394	0.433	0.539
36	0.279	0.330	0.388	0.427	0.533
37	0.275	0.325	0.383	0.421	0.526
38	0.271	0.321	0.378	0.415	0.519
39	0.267	0.317	0.373	0.410	0.513
40	0.264	0.313	0.368	0.405	0.507
50	0.235	0.279	0.329	0.363	0.456
60	0.214	0.255	0.300	0.331	0.418
70	0.198	0.235	0.278	0.307	0.388
80	0.185	0.220	0.260	0.287	0.363
90	0.174	0.207	0.245	0.271	0.343
100	0.165	0.197	0.233	0.257	0.326

(From Zar, 1974.)

Index